INFRARED THERMOGRAPHY
for Geomechanical Model Test

Gong Weili
Wang Jiong
Liu Dongqiao

Science Press
Beijing

INTRODUCTION

The book starts, in Chapter 1, with an overview of the current status of geomechanical model tests and infrared detection, followed by an introduction to some theoretical aspects of infrared physics and algorithms used in the infrared image processing (Chapter 2). Geomechanical model construction is presented in Chapter 3. In Chapters 4, 5, 6, 7 and 8, application of infrared thermography in the geomechanical model tests on tunnel excavations and roadway stability assessments in differently inclined rock strata are presented respectively with a detailed description of the experimental methods, testing procedures, imaging processing algorithms and findings obtained from characterization of the thermal sequences.

The intended readers may fall into three groups. The first group is the undergraduate science major students who may want to learn in depth about the topic of this book. In this case, the publication can complement an advanced text book. The second may be the engineers who will find valuable information about the practical application of the TNDE technique. The last group includes the graduate students, Ph. D students and researchers who will find the systematic introduction of the principles, methods and algorithms for construction of the geomechanical models, performing the infrared detection, processing and characterization of the obtained thermal images.

图书在版编目(CIP)数据

红外热成像技术及其岩土实验应用＝Infrared Thermography for Geomechanical Model Test:英文/宫伟力,王炯,刘冬桥著.—北京:科学出版社,2016.1

ISBN 978-7-03-044777-7

Ⅰ.①红… Ⅱ.①宫… ②王… ③刘… Ⅲ.①红外成像系统-应用-岩土工程-实验-英文 Ⅳ.①TU4-33

中国版本图书馆 CIP 数据核字(2015)第 122346 号

责任编辑:李 雪/责任校对:桂伟利
责任印制:徐晓晨/封面设计:耕者设计工作室

科学出版社出版
北京东黄城根北街 16 号
邮政编码:100717
http://www.sciencep.com

北京厚诚则铭印刷有限公司 印刷
科学出版社发行 各地新华书店经销

*

2016 年 1 月第 一 版 开本:720×1000 B5
2017 年 1 月第二次印刷 印张:17 7/8 插页:8
字数:360 000
定价:128.00元
(如有印装质量问题,我社负责调换)

Infrared Thermography for Geomechanical Model Test

Gong Weili, Ph. D. , Professor
Associate Professor of Fluid Mechanics and Rock Dynamics
China University of Mining and Technology
Beijing China

Wang Jiong, Ph. D. , Lecturer
Senior Lecturer in Engineering Rock Mechanics
China University of Mining and Technology
Beijing China

Liu Dongqiao, Ph. D. , Lecturer
Senior Lecturer in Rock Mechanics and Geotechnical Engineering
China University of Mining and Technology
Beijing China

About the authors

Gong Weili, Dr. & Professor

Weili Gong obtained his Ph. D. in 1999 at the University of Science and Technology Beijing (USTB). He was working as a postdoctoral fellow on engineering mechanics from 1999 to 2001 at China University of Mining and Technology Beijing (CUMTB) and spent his professional career in fluid mechanics at School of Mechanics and Civil Engineering, CUMTB, as associate professor from 2001-2013—in teaching and research.

From 2013, he has been working as senior research fellow in rock dynamics at State Key Laboratory for Geomechanics and Deep Underground Engineering. His personal research interests involve fluid mechanics, rock dynamics, and thermography-based nondestructive evaluation (TNDE). He has written over 90 scientific papers and published two books on these subjects.

Wang Jiong, Dr. & Lecture

Jiong Wang obtained his Ph. D. in 2011 at the China University of Mining and Technology Beijing (CUMTB). He has been working as a senior research fellow on rock mechanics at State Key Laboratory for Geomechanics and Deep Underground Engineering and a lecturer on engineering mechanics at the School of Mechanics and Civil Engineering, CUMTB, from 2011 till now.

His personal research interests involve geotechnical engineering, rock mechanics, and TNDE. He has written over 30 scientific papers and published one book on these subjects.

Liu Dongqiao, Dr. & Lecture

Dongqiao Liu obtained his Ph. D. in 2014 at the China University of Mining and Technology Beijing (CUMTB). From 2014, he is working as a principal investigator at State Key Laboratory for Geomechanics and Deep Underground Engineering, CUMTB.

His personal research interests involve rock mechanics and geotechnical engineering. He has written over 20 scientific papers on these subjects.

Forward

The object for rock mechanics research in lithosphere is naturally created and composed mainly by the geological bodies. Different from most of the man-made systems, the geoscientific system has not been fully understood by human beings so far due to its complexity and non-transparency. As the mining proceeds to deeper ground, the requirement for safety mining and disaster mitigation imposes great challenges to the community of rock mechanics.

The use of infrared thermography for detecting large-scale geomechanical model tests is a good attempt to unveil fundamental features of nonlinearity and heterogeneity of the deep-buried underground structures in complex geological conditions. The thermal inspection work conducted by the authors began in 2007 and significant improvement in understanding of the rock behaviours in terms of the thermomechanical coupling has been achieved.

I had the pleasure and privilege of reading the original book, especially some important chapters of this book regarding the application of infrared thermography to the large-scale geomechanical model tests carried out in the state-key laboratory for geomechanics and deep underground engineering (SKL-GDUE) at China University of Mining and Technology Beijing (CUMTB), and think that they are very interesting and meaningful to the readers involved in the related areas

A marked difficulty encountered in application of the thermography in geomechanics is the problem of the low-contrast thermography as a result of the large-scale imaging area and room temperature influences. Except for the distinctive thermal responses of rock masses, the book presents the authors's expertise in processing and analyzing thermograms acquired in the large-scale geomechanical model tests. I hope it will contribute to the interested readers to get some implications for coping with the challenges said in the first paragraph.

He Manchao
Academician of CAS, Professor, Ph. D.
Beijing, 2015

Preface

Infrared radiation emitted by the stressed materials carries specific information involving the deformed configuration at different scales and the heat patterns produced by the intrinsic dissipation of energy caused by friction between grains due to dislocation and deformation at microscopic scales. This information can be remotely sensed by infrared thermography in real time and without any contact or intrusion to the materials under investigation. The infrared thermography-based monitoring technique is often referred to as nondestructive evaluation (TNDE).

Unlike visible spectra images (wavelength spectrum: 0.35-0.75 μm) which are produced by reflection and reflectivity differences, infrared images (wavelength spectrum: 0.75-100 μm) are produced by a self-emission phenomenon and also by variations in emissivity. Thermal image represents rock response based on the thermal-mechanical coupling and does not require supplementary lighting as well. When processed with proper algorithms, thermal image will not only be able to detect geometrical features such as crack propagation, but also the static and dynamic friction which could hardly be observed by the conventional optical visualization techniques.

The potential of nondestructive evaluation of materials by infrared thermography is being exploited, especially since the availability, in the 1980s (Reynolds, 1988), of commercial infrared cameras whose video signals are compatible with black-and-white television standards (Xavier, 1993). Since the 1990s, infrared thermography has gained widespread recognitions as a NDE technique and widely been applied, but not limited, to detecting damage in the deformed materials such as composite (Connolly and Copley, 1990), carbon fiber reinforced polymers (Steinberger et al., 2006), metals (Luong, 1995; Pastor et al., 2008), concrete and rock (Brady and Rowell, 1986; Luong, 1990, 2007; Grinzato et al., 2004; Wu et al., 2006; Shi et al., 2007), high-speed water jets (Gong et al., 2008).

In recent years, the use of the infrared thermography has been extended to the geomechanical model tests involving tunnel excavations (He et al., 2010a, 2010b; He, 2011; Gong et al., 2013) and roadway stability assessment under great overburdens (He et al., 2009; Gong et al., 2015a, 2015b). The size of geomechanical models is usually larger than that of the laboratory specimens and the obtained thermal image in detecting the geomechanical models will have smaller spatial resolution. Meanwhile, the simulated rock masses in the geomechanical model tests often comprise geological discontinuities. As a result, infrared thermography should be working at passive mode due to the low and nonuniform thermal conductivity.

The active thermography (i. e. the thermography working at active mode) uses a known external heat source for heating up the tested object and, therefore, the obtained temperature needs to calibrate and increases physically with the heating source. As the heating process is operated in a controlled way, a large temperature increment of the stressed materials could be obtained. The active thermography is usually employed in the case of the materials with a relatively simple constitutive relationship such as metals and composites. When employing the passive thermography in detecting a large-scale object such as the geological models, temperature increment may be much smaller and the resultant thermal image is often vague in contrast. In this case, advanced imaging processing is required.

With this book, we want to present the reader with our work over years in application of the infrared thermography in geomechanical model tests. Our expertise in processing the low-contrast and noisy thermal image obtained in detecting the large-scale object with the thermography working at passive mode will be introduced. Emphasis is also given in construction of the geological models and our research findings in which some readers may find surprising the many departures from the knowledge obtained using the conventional detecting methods or numerical simulations of the problems under the comparable situation.

The intended readers may fall into three groups. The first group is the undergraduate science major students who may want to learn in depth about the topic of this book. In this case, the publication can complement an advanced text book. The second may be the engineers who will find valuable information about the practical application of the TNDE technique. The last group includes the graduate students, Ph. D. students and researchers who will find the systematic introduction of the principles, methods and algorithms for construction of the geomechanical models, performing the infrared detection, processing and characterization of the obtained thermal images.

The book starts, in Chapter 1, with an overview of the current status of the geomechanical model tests and infrared detection, followed by an introduction to some theoretical aspects of the infrared physics and the algorithms used in the infrared image processing (Chapter 2). Geomechanical model construction is presented in Chapter 3. In Chapters 4, 5, 6, 7 and 8, application of the infrared thermography in the geomechanical model tests on tunnel excavations and roadway stability assessments in differently inclined stratified rock strata are reported respectively with the detailed description of the experimental methods, testing procedures, imaging processing algorithms and the findings obtained from the characterization of the thermal sequences.

Financial support from the special funds for the Major State Basic Research Project under Grant No. 2006CB202200, the Innovative Team Development Project of the State Educational Ministry of China under Grant No. IRT0605, the National Natural Science Foundation of China (key project) under Grant No. 51134005, the National Natural Science Foundation of China under Grant No. 51574248 and 51404278 are gratefully acknowledged.

Chengrong Ma, associate Professor in Shaoxing University is appreciated for his contribution to this book in processing the photographs in Chapter 2. From 2007, many collaborators with the authors of this book participated in the geomechanical model tests, including Liu Zhe, Zhao Yuan, Zhai Huiming, Wang Hongjian, graduate students when the work ongoing; and Haipeng Zhang, laboratory assistant. Their contributions to this book are acknowledged. Thanks should also be given to Dr. Yanyan Peng, now a lecturer in Shaoxing University and having been a Ph. D. candidate in our institute, for her editorial work to this book.

Last but not the least, the authors of this book would like to expresses their heartfelt thanks to Prof. Manchao He, academician of the Chinese Academy of Sciences, who is the chief scientist for the three state-level major scientific research projects mentioned previously which are the principal sources of the financial support to this work, for his first introduction of infrared thermography to the geomelchanics experiments. Besides, Prof. He developed the frame for conducting the geomechanical model test, designed all the geomechanical model tests and proposed the original ideas and mechanics models used in this book. This book would not have been possible without Prof. He's great contributions.

Gong Weili
Beijing, 2015

Contents

Preface

Chapter 1 Overview 1
1.1 Background 1
1.2 Overview of geomechanics model tests 3
1.3 Overview of infrared detection 4
References 6

Chapter 2 Theoretical aspects of the infrared 11
2.1 The infrared 11
2.2 Infrared spectral band 13
2.3 Radiometry fundamentals 18
    2.3.1 Radiant energy 19
    2.3.2 Radiant power and flux 19
    2.3.3 Geometrical spreading of a beam 19
    2.3.4 Radiance 20
    2.3.5 Irradiance 22
    2.3.6 Radiant exitance 22
    2.3.7 Radiant intensity of a source in a given direction ... 23
    2.3.8 Bouguer's law 23
    2.3.9 Radiation scattering 24
2.4 Black body radiation 25
    2.4.1 Concept of black body 25
    2.4.2 Planck's law 26
    2.4.3 Wien's law 29
    2.4.4 Stefan-Boltzmann law 30
    2.4.5 Exitance of a black body in a given spectral band ... 30
    2.4.6 Calculation of exitance of black body 32
    2.4.7 Thermal radiation contrast 35
2.5 Radiation of real bodies 36
    2.5.1 Different types of radiator 36

2.5.2		Emissivity of a material	38
2.5.3		Stefan-Boltzmann's law for grey body	43
2.5.4		Dielectric materials	44
2.5.5		Electrically conducting materials	47

References 48

Chapter 3　Geomechanical model test 50

3.1　Literature review on physical model test 50
3.2　Similarity theory and dimensional analysis 53
　　3.2.1　Similarity principles 53
　　3.2.2　Selection of similarity materials and ratios 56
3.3　Field case (prototype) 58
　　3.3.1　Site geology 58
　　3.3.2　In situ rock properties 60
3.4　Geomechanical model construction 61
　　3.4.1　Testing machine 61
　　3.4.2　Model dimension 62
　　3.4.3　Physico-mechanical parameters of the model 64
　　3.4.4　Rock structure simulation 65
　　3.4.5　Geomechanical model 68
3.5　Infrared detection 69
　　3.5.1　Thermography and imaging procedures 69
　　3.5.2　Temperature calibration 72
　　3.5.3　Image processing 73

References 74

Chapter 4　Excavation in 60° inclined strata 78

4.1　Introduction 78
4.2　Experiment 79
　　4.2.1　Rock model material 79
　　4.2.2　Geomechanical model construction 80
　　4.2.3　Excavation plan 82
　　4.2.4　Excavation method 83
4.3　Infrared detection 84
　　4.3.1　Infrared thermography 84
　　4.3.2　Thermal-mechanical coupling 85
4.4　Image processing 87

4.4.1	Problem statement	87
4.4.2	Algorithms	87
4.4.3	Processing and assessment	90

4.5 Image analysis ··· 91
    4.5.1 Extracting the energy release index ··· 91
    4.5.2 Spectral characterization ··· 92
    4.5.3 Principles for image analysis ··· 94

4.6 Experimental results ··· 94
    4.6.1 Overall thermal response ··· 94
    4.6.2 Heat sources and thermal conduction ··· 96
    4.6.3 Characterization of the full-face excavation ··· 96
    4.6.4 Heat production mechanism in the staged excavation ··· 103
    4.6.5 Characterization of the staged excavation ··· 104

4.7 Discussion ··· 112
    4.7.1 Excavation in differently inclined rocks over full-face excavation ··· 112
    4.7.2 Excavation in differently inclined rocks over the staged excavation ··· 115
    4.7.3 Summary ··· 119

References ··· 120

Chapter 5 Excavation in 45° strata ··· 123

5.1 Introduction ··· 123
5.2 Short review of infrared detection ··· 124
5.3 Experiment ··· 127
    5.3.1 Model construction ··· 127
    5.3.2 Testing procedure ··· 129
5.4 Infrared detection ··· 131
    5.4.1 Infrared thermography ··· 131
    5.4.2 Energy release index ··· 132
    5.4.3 Image processing algorithm ··· 133
    5.4.4 Principles for image analysis ··· 133
    5.4.5 Fourier analysis ··· 133
5.5 Results and Discussions ··· 135
    5.5.1 Overall thermal response ··· 135
    5.5.2 Characterization of the full-face excavation ··· 138

5.5.3　Comparison between the excavation in 0° and 45° inclined strata ……… 143
　　　5.5.4　Characterization of the staged excavation ……… 145
　　　5.5.5　Summary ……… 152
References ……… 153
Chapter 6　Excavation in horizontal strata ……… 156
6.1　Introduction ……… 156
6.2　Experiment ……… 158
　　　6.2.1　Geomechanical model construction ……… 158
　　　6.2.2　Testing procedure ……… 159
6.3　Infrared detection ……… 161
　　　6.3.1　Infrared thermography ……… 161
　　　6.3.2　Image processing ……… 162
　　　6.3.3　Fourier transform of the thermal image ……… 163
　　　6.3.4　Enhancement of the thermal image ……… 164
　　　6.3.5　Spectral analysis ……… 166
6.4　Results and discussions ……… 167
　　　6.4.1　Overall thermal response ……… 167
　　　6.4.2　Characterization of the full-face excavation ……… 169
　　　6.4.3　Characterization of the staged excavation ……… 175
　　　6.4.4　Summary ……… 181
References ……… 183
Chapter 7　Overloaded tunnel in 45° inclined rocks ……… 185
7.1　Introduction ……… 185
7.2　Experimental ……… 187
　　　7.2.1　Geomechanical model ……… 187
　　　7.2.2　Loading path ……… 188
7.3　Infrared detection ……… 189
　　　7.3.1　Infrared thermography and imaging procedures ……… 189
　　　7.3.2　Temperature calibration ……… 191
　　　7.3.3　Image processing ……… 192
7.4　Fourier analysis ……… 193
　　　7.4.1　Stress wave propagation ……… 193
　　　7.4.2　Fourier transform ……… 196
　　　7.4.3　Periodicity in time domain ……… 197

	7.4.4	Periodicity in spatial domain	199
	7.4.5	Physical meaning of the spatial frequency	199
	7.4.6	Method for spectral analysis	200
7.5	Loading path and overall rock response	203
	7.5.1	Energy release index	203
	7.5.2	Loading rate	205
	7.5.3	Characterization of the loading rate effect	205
7.6	Results and discussions	207
	7.6.1	Terms and approach	207
	7.6.2	Spectra characterization of loading state A	207
	7.6.3	Characterization of loading cases with slow loading rate	212
	7.6.4	Characterization of loading cases with fast loading rate	218
	7.6.5	Discussions	222
	7.6.6	Summary	225
References	226

Chapter 8 Overloaded tunnel in horizontal strata — 231

8.1	Introduction	231
8.2	Experimental	232
	8.2.1	Geological model	232
	8.2.2	Loading scheme	233
	8.2.3	Infrared detection	234
8.3	Problem statement	236
8.4	Image denoising filters	237
	8.4.1	Types of the noise	237
	8.4.2	Removing environmental noise	237
	8.4.3	Suppression of the impulsive noise	238
	8.4.4	Removing the additive-periodical noise	238
8.5	Morphological enhancement filter	239
	8.5.1	Short review	239
	8.5.2	Fundamentals	239
	8.5.3	Filter development	240
	8.5.4	Multi-scale SE	242
8.6	Image processing	243
	8.6.1	Algorithm and image analysis rules	243
	8.6.2	Assessment of imaging processing effect	244

8.6.3	Rock response at hydrostatic stress state	245
8.6.4	Rock response at unbalanced stress state	249
8.7	Characterization of new IR images	253
8.7.1	Mission and rule	253
8.7.2	Loading case B1	255
8.7.3	Loading case B2	256
8.7.4	Loading case B3	257
8.7.5	Loading case B4	259
8.7.6	Loading case B5	261
8.7.7	Loading case B6	262
8.7.8	Discussion	264
8.7.9	Summary	266
References		266
Appendix: The colered thermal images in chapter 4-8		269

Chapter 1　Overview

1.1　Background

With fast growing demands for natural resources, more and more deep-seated tunnels and openings are being planned and constructed (Sagong and Bobet, 2002). Understanding the mechanism of rock damage in the creation and operation of an underground cavern in jointed rock masses has been a topic of research for engineers in various fields including, for example, developing transportation tunnels, petroleum production drillings, nuclear waste disposal, high-head diversion hydropower stations and underground mines (He et al., 2005; Zhao et al., 2007; Zhu et al., 2011; Lin et al., 2015).

For example, the mining depth of the underground coal mines in China reached at present as deep as 1070 m and will proceed below 1500 m in the near future (He et al., 2002); to the year of 2010, 3000 m deep underground mines account for 30% in South Africa (Malan and Basson, 1998); in the New Alpine Transverse project, the Gotthard Base Tunnel consists of two parallel single track tubes, each with a diameter of 9.2 m and has a length of 57 m and a maximum overburden of 2350 m (Hagedorn and Stadelmann, 2008); with a cover depth of 1400 m, the 22-km-long Qinling tunnel is cut across the Qinling mountain range to link up the cities of Xi'an and Ankang, China (Lee et al., 1996); in the Jinping Ⅱ hydropower project, four parallel power tunnels, each with a diameter of 12-13 m and a length of about 17 km, are excavated beneath the Jinping mountain at a depth of between 1500 and 2000 m with a maximum depth of about 2525 m (Lin et al., 2015).

Engineering rock masses can be classified into three categories: ①muddy rock and mud or calcareous cemented sand rock; ②iron or silica cemented sand rock and crystalline rock; ③rock masses with structural planes. Muddy rocks, mud or calcareous cemented sand rocks at shallow depth deforms by swelling due to the adsorption of water and by squeezing due to the great overburden. Iron or silica cemented sand rock and crystalline rock does not deform under shallow depth and fails by spalling or bursting under great depth. Failure mode for the structural plane-contained rock masses under great depth is the large deformation by shear on the structural planes (He et al., 2002). It was pointed out by He et al. (2005), *"nonlinearity and complexity are the principal difficulties encountered at great depth, some conventional theories and constitutive relations for linear rock mechanics do not work, and research on basic theories for deep rocks mechanics is, therefore, imperative"*.

The *"complexities"* come from the geological and mechanical conditions for rock mechanics at great depth. It can be summarized as the *"three highs plus one perturbation"*, i. e. *"high in-situ stresses, high geothermal temperature, high pore pressure, as well as the disturbance from mining activities"* (He et al., 2005). One of the major challenges facing in the deep rock mechanics is the safety evaluation of the tunnels or caverns in construction and operation. The significant difficulty encountered in the safety evaluation is the understanding of the nonlinearity of the structural responses of deep-buried tunnels. Thus, it is essential to further develop various models and analysis methods for modernization of the tunnel safety design, which requires studying the cracking and failure of the surrounding rocks under excavation and the stability of the tunnel under various loading cases (Lin et al., 2015).

Extensive researches have been conducted on tunneling, roadway excavation and reinforcement, block caving and stability of the underground caverns at depths including, for example, in-situ tests (Read, 2004; Li et al., 2008); analytical studies (Lydzba et al., 2003); numerical modeling using finite difference method (FEM) (Tsesarsky, 2012), finite element method (FEM) (Golshani et al., 2007; Fortsakis et al., 2012), discrete element method (DEM) (Heuze and Morris, 2007), discontinuous deformation analysis (DDA) (Hatzor and Benary, 1998; Tsesarsky and Hatzor, 2006; Mazor et al., 2009; Zuo et al., 2009), and physical model tests (Sharma et al., 2001; Kamata and Masimo, 2003; Liu et al., 2003; Castro et al., 2007; Lee and Schubert, 2008; Shin et al., 2008; Fekete et al., 2010; Zhu et al., 2011; Li et al., 2013). Well designed experiments and judicious choice of model materials with the aid of data acquisition methods such as infrared thermography may yield important views into failure modes and mechanisms that are not available from numerical models (Zhu et al., 2011), constituting the topics documented in this book.

1.2 Overview of geomechanics model tests

Geomechanical model test is one of the important means of the physical model test for studying the stability of the large-scale geotechnical engineering structure in hydraulics (Li et al., 2005; Zhu et al., 2010; Lin et al., 2013, 2015), transportation (Meguid et al., 2008; Huang et al., 2013) and mining (He et al., 2010a, 2010b; He, 2011; Gong et al., 2015a, 2015b). By utilizing dimensional analysis and the similarity principles, the geomechanical model test is capable of simulating engineering rock masses containing geological discontinuities such as bedding planes, joints and faults; at the same time, the geomechanical model test can also be used to study the progressive development of the excavation damaged zone (EDZ) and the failure process for tunnels and underground caverns under great overburdens.

Extensive studies on the geomechanical model tests have been conducted on the tunnel excavation and operation. As reviewed by Meguid et al. (2008), the conventional physical model tests includes, for example, 2D trap door tests (Jeon et al., 2004), rigid tubes with flexible or rigid face tests (Kamata and Masimo, 2003), a miniature tunnel boring machine (TBM) and other techniques (Nomoto et al., 1999). These reported studies concerning physical modeling largely involved small scale model tests. The limitations of small scale model tests concern the inability of faithful simulation of in situ stresses and inconsistencies in scaling factors for different variables (e.g., length, inertia force, creep, etc.). To overcome these problems, large-scale geomechanical model test is required, such as those conducted by Lee and Schubert (2008), Sharma et al. (2001), Kamata and Masimo (2003), Zhu et al. (2011), Fekete et al. (2010), Liu et al. (2003), Castro et al. (2007), Shin et al. (2008) and Li et al. (2008), which will be discussed in this book.

In recent decades, some remote sensing techniques are incorporated into the geomechanical tests in order to capture the geotechnical information about detailed rock conditions and responses. Laser scanning and photogrammetry are two imaging techniques widely used in a tunnel environment. Digital imaging system for determining displacement and strain has been applied in recent decades to a number of geotechnical engineering problems (Lee and Bassett, 2006; Birch, 2008; Gaich and Potsch, 2008), and recently used successfully to measure convergence around cavern in large-scale three-dimensional geomechanical model tests (Zhu et al., 2010, 2011).

The photogrammetry, as reviewed by Fekete et al. (2010), requires supplementary lighting while three-dimensional laser scanning (Lidar) acts as its own source of "illumination" (Kim et al., 2006; Birch, 2008). The Lidar was applied but not limited to the evaluation of rock reinforcement (Gosliga et al., 2006), landslide monitoring (Strouth and Eberhardt, 2005), and stratigraphy modeling (Buckley et al., 2008). Fekete et al. (2010) used improved Lidar in active tunneling environment under dusty, damp, and dark conditions and collected very accurate, high resolution 3-dimensional images of its surroundings.

The advantages of employing non-contact optical vision techniques lie in their ability to represent the structural change by realistic and practical surface models or geometrical features. Usability of the detected geometrical features such as cracks and discontinuities, however, depends on the image resolution and does not have a definite relation to stress redistribution in the surrounding rocks. Infrared (IR) thermography is another non-contact and remote sensing technique which produces thermal image in real time by detecting electromagnetic waves within infrared wave band (Luong, 1995). Thermal image represents rock response based on the thermal-mechanical coupling effect (Luong, 2007) and does not require supplementary lighting as well.

When processed with proper algorithms, thermal image will not only be able to detect geometrical features such as crack propagation, but also the static and dynamic friction (He et al., 2010a) which could hardly be observed by the conventional optical visualization techniques. Thermography matrix data set is in fact the infrared radiation temperature field on the surface in view induced by energy release of the cracking rocks. The fact that frequency-spectra of the thermal image can represent the seismic wave propagation is the intrinsic advantage of the thermography that the optical imaging technique does not possess. Principles, techniques and the application of the infrared thermography in geomechanics model test will be discussed in this book.

1.3 Overview of infrared detection

Although some contact detection technologies, such as acceleration sensors, strain gauges, stress wave propagation meters and displacement measurement, have been successfully applied both in situ and laboratory, but they are not suited for such dynamical processes as rock impaction because of the intrusive disturbance and damage-prone problem of the contacting detectors (Shi et al., 2007). The widely used non-destructive detection technique in rock mechanics includes acoustic emission (AE) and electromagnetic radiation. It describes the progressive development failure of the rock indirectly by the transformation or statistics of the test data sets (Luo et al., 2006).

Owing to the thermomechanical coupling, infrared thermography provides a non-destructive, non contact and real-time test to observe the physical process of material degradation and to detect the occurrence of intrinsic dissipation without surface contact or in any way influencing the actual surface temperature of the tested object (Luong, 1995). It produces heat images directly from the invisible radiant energy (dissipated energy) emitted from stationary or moving objects at any distance. By using an infrared thermographic detection system, one can observe and capture the physical and structural changes of an object, represented by the infrared radiation temperature changes, in real time and over the entire field of the viewed surface.

The relationship between dissipated energy and the tested objects temperature obeys the Stefan-Boltzmann law:

$$M = \varepsilon \sigma T^4 \qquad (1.1)$$

where, M is the radiant exitance, W/m^2; ε is the object's emissivity, $0 < \varepsilon < 1$; σ is the Stefan-Boltzmann constant, 5.67×10^{-8}, $J/(m^2 \cdot K^4)$; and T is the absolute temperature.

Abnormal rising of the surface radiation temperature were observed on the infrared thermal images in the climate satellites several days before and after earthquake. It was recognized that the earthquake is related to the change of the stress field of the earth's crust, and then the change of the stress field causes electromagnetic radiation including the infrared radiation which has the thermal effect (Qiang et al., 1990).

The fact that infrared radiation energy varies with the change of stress field of the loaded rocks was further verified by the subsequent studies: uniaxial and biaxial loading of sedimentary rocks and igneous rocks (Cui et al., 1993; Zhi et al., 1996); the structural stress fields of rock and gas burst disasters resulting in local abnormal high temperature in mine (Wu and Wang, 1998); projectile impact on rock (Shi et al., 2007); the roadway tunnel under plane loading (He et al., 2009) and its excavation process (He et al., 2010a, 2010b) in the physical analogue models in a horizontal and vertical strata consisting of alternating layers of sandstone, mudstone and coal.

Infrared thermography has been widely applied, but not limited to, in detecting damage in the deformed materials such as composite (Connolly and Copley, 1990), carbon fiber reinforced polymers (Steinberger et al., 2006), metals (Luong, 1995; Pastor et al., 2008), and rocks and rock-like materials (Luong, 1990, 2007; Qiang et al., 1990; Cui et al., 1993; Zhi et al., 1996; Geng et al., 1998; Wu and Wang, 1998; Wu et al., 2002, 2004, 2006; Brady and Rowell, 2004; Grinzato et al., 2004; Shi et al., 2007). In recent decades, infrared thermography has been employed to detect the excavation damaged zone (EDZ) in the large-scale geomechanical model tests for simulation of tunnel excavations (He et al., 2010a, 2010b; He, 2011; Gong et al., 2013a, 2013b, 2015a, 2015b).

References

Birch J S. 2008. Using 3D analyst mine mapping suite for underground mapping. In: Tonon F (ed) Proceedings of Laser and Photogrammetric Methods for Rock Tunnel Characterization Workshop. 42nd US Rock Mechanics Symposium ARMA, San Francisco, June 28-29, Cdrom.

Brady B H G, Brown E T. 2004. Rock Mechanics for Underground Mining. New York: Kluwer Academic Publishers.

Buckley S, Howell J, Enge H, et al. 2008. Terrestrial laser scanning in geology: data acquisition, processing and accuracy considerations. Journal of the Geological Society, London 165 (3): 625-638.

Castro R, Trueman R, Halim A. 2007. A study of isolated draw zones in block caving mines by means of a large 3D physical model. International Journal of Rock Mechanics and Mining Sciences, 44(6): 860-870.

Connolly M, Copley D. 1990. Thermographic inspection of composite material. Materials Evaluation, 48,(12): 1461-1463.

Cui C, Deng M, Geng N. 1993. Study on the features of spectrum radiation of rocks under different load. Chinese Science. Bulletin, 38(6): 538-541.

Fekete S, Diederichs M, Lato M. 2010. Geotechnical and operational applications for 3-dimensional laser scanning in drill and blast tunnels. Tunnelling and Underground Space Technology, 25: 614-628.

Fortsakis P, Nikas K, Marinos V, et al. 2012. Anisotropic behavior of stratified rock masses in tunneling. Engineering Geology, 141-142: 74-83.

Gaich A, Potsch M. 2008. Computer vision for rock mass characterization in underground excavations. In: Tonon F (ed) Proc. Laser and Photogrammetric Methods for Rock Tunnel Characterization. 42nd US Rock Mechanics Symposium ARMA, San Francisco, June 28-29.

Geng N G, Yu P, Deng M D, et al. 1998. The simulated experimental studies on cause of thermal infrared precursor of earthquake. Earthquake, 18: 83-88.

Golshani A, Oda M, Okui Y, et al. 2007. Numerical simulation of the excavation damaged zone around an opening in brittle rock. International Journal of Rock Mechanics and Mining Sciences, 44: 835-845.

Gong W L, Gong Y X, Long Y F. 2013a. Multi-filter analysis of infrared images from the excavation experiment in horizontally stratified rock. Infrared Physics and Technology, 56: 57-68.

Gong W L, Peng Y Y, He M C, et al. 2015b. Thermal image and spectral characterization of roadway failure process in geologically 45° inclined rocks. Tunnelling and Underground Space Technology, 49: 156-173.

Gong W L, Peng Y Y, Sun X M, et al. 2015a. Enhancement of low-contrast thermograms for detecting the stressed tunnel in horizontally stratified rocks. International Journal of Rock Mechanics and Mining Sciences, 74: 69-80.

Gong W L, Wang J, Gong Y X, et al. 2013b. Thermography analysis of a roadway excavation experiment in 60° inclined stratified rocks. International Journal of Rock Mechanics and Mining Sciences, 60: 134-147.

Gosliga F V, Lindenbergn R, Pfeifer N. 2006. Deformation analysis of a bored tunnel by means of terrestrial laser scanning. In: Image Engineering and Vision Metrology. ISPRS Commission, 36: 167-172.

Grinzato E, Marinetti S, Bison P G, et al. 2004. Comparison of ultrasonic velocity and IR thermography for the characterization of stones. Infrared Phys. &. Technol., 46: 63-68.

Hagedorn H, Stadelmann R. 2008. Gotthard base tunnel rock burst phenomena in a fault zone, measuring and modeling results. In: Proceedings of Underground Facilities for Better Environment and Safety. World Tunnel Congress, India.

Hatzor Y H, Benary R. 1998. The stability of a laminated Voussoir beam: Back analysis of a historic roof collapse using DDA. International Journal of Rock Mechanics and Mining Sciences, 2(12): 165-181.

He M. C. 2011. Physical modeling of an underground roadway excavation in geologically 45° inclined rock using infrared thermography. Engineering Geology, 121(3-4): 165-176.

He M C, Gong W L, Li D J, et al. 2009. Physical medeling of failure process of the excavation in horizontal strata based on IR thermography. Mining Science and Technology, 19(6): 689-698.

He M C, Gong W L, Zhai H M, et al. 2010a. Physical modeling of deep ground excavation in geologically horizontally strata based on infrared thermography. Tunnelling and Underground

Space Technology Technol. 25: 366-376.

He M C, Jia X N, Gong W L, et al. 2010b. Physical modeling of an underground roadway excavation vertically stratified rock using infrared thermography. International Journal of Rock Mechanics and Mining Sciences, 47: 1212-1221.

He M C, Lu X J, Jing H H. 2002. Characters of surrounding rockmass in deep engineering and its non-linear dynamic-mechanical design concept. Chinese Journal of Rock Mechanics and Engineering, 21(8): 1215-1224.

He M C, Xie H P, Peng S P, et al. 2005. Study on rock mechanics in deep mining engineering. Chin. J. Rock Mech. Eng. , 24(16): 2803-2813.

Heuze F E, Morris J P. 2007. Insights into ground shock in jointed rocks and the response of structures there-in. International Journal of Rock Mechanics and Mining Sciences, 44: 647-676.

Huang F, Zhu H H, Xu Q W, et al. 2013. The effect of weak interlayer on the failure pattern of rock mass around tunnel-scaled model tests and numerical analysis. Tunnelling and Underground Space Technology, 35: 207-218.

Jeon S, Kim J S, Hong Y C. 2004. Effect of a fault and weak plane on the stability of a tunnel in rock-a scaled model test and numerical analysis. International Journal of Rock Mechanics and Mining Science, 41(1): 658-663.

Kamata H, Masimo H. 2003. Centrifuge model test of tunnel face reinforcement by bolting. Tunnelling and Underground Space Technology, 18(2): 205-205.

Kim C, Ghanma M, Habib A. 2006. Integration of photogrammetric and LIDAR data for realistic 3-dimensional model generation. In: 1st International Workshop on Mobile Geospatial Augmented Reality, Banff, Alberta, Ganada, May 29-30, Cdrom.

Lee C F, Wang S J, Yang Z F. 1996. Geotechnical aspects of rock tunneling in China. Tunnelling and Underground Space Technology, 11(4): 445-454.

Lee Y J, Bassett R H. 2006. Application of a photogrammetric technique to a model tunnel. Tunnelling and Underground Space Technology, 21(1): 79-65.

Lee Y Z, Schubert W. 2008. Determination of the length for tunnel excavation in weak rock. Tunnelling and Underground Space Technology, 23, 221-231.

Li S C, Hu C, Li L P, et al. 2013. Bidirectional construction process mechanics for tunnels in dipping layered formation. Tunnelling and Underground Space Technology, 36: 57-65.

Li S J, Yu H, Liu Y X, et al. 2008. Results from in situ monitoring of displacement, bolt load, and disturbed zone of a power house cavern during excavation process. International Journal of Rock Mechanics and Mining Sciences, 45: 1519-1525.

Li Z K, Liu H, Dai R, et al. 2005. Application of numerical analysis principles and key technology for high fidelity simulation to 3-D physical model tests for underground caverns. Tunnelling and Underground Space Technology, 20(4): 390-399.

Lin P, Liu H Y, Zhou W Y. 2015. Experimental study on failure behaviour of deep tunnels under high in-situ stresses. Tunnelling and Underground Space Technology, 46: 28-45.

Lin P, Zhou Y N, Liu H Y, et al. 2013. Reinforcement design and stability analysis for large-span tailrace bifurcated tunnels with irregular geometry. Tunnelling and Underground Space Technology, 38(9): 189-204.

Liu J, Feng X T, Ding X L, et al. 2003. Stability assessment of the Three-gorges Dam foundation, China using physical and numerical modeling-part Ⅰ. Physical model tests. International Journal of Rock Mechanics and Mining Sciences, 40(5): 609-631.

Long M P. 1995. Infrared thermographic scanning of fatigue in metals. Nuclear Engineering and Design, 158: 363-376.

Luo X, Haya H, Inaba T, et al. 2006. Seismic diagnosis of railway substructures by using secondary acoustic emission. Soil Dynamics and Earthquake Engineering, 26(12): 1101-1110.

Luong M P. 1990. Infrared thermovision of damage processes in concrete and rock. Engineering Fracture Mechanics, 35: 127-135.

Luong M P. 2007. Introducing infrared thermography in soil dynamics. Infrared Phys. &. Technol. , 49, 306-311.

Lydzba D, Pietruszczak S, Shao J F. 2003. On anisotropy of stratified rocks, homogenization and fabric tensor approach. Computers and Geotechnics, 30: 289-302.

Malan D F, Basson F R P. 1998. Ultra-deep mining: the increased potential for squeezing conditions. J. S. Afr. Inst. Min. Metall, 98(11/12): 353-363.

Mazor D B, Hatzor Y H, Dershowitz W S. 2009. Modeling mechanical layering effects on stability of underground openings in jointed sedimentary rocks. International Journal of Rock Mechanics and Mining Sciences, 46: 262-271.

Meguid M A, Saada O, Nunes M A, et al. 2008. Physical modeling of tunnels in soft ground: a review. Tunnelling and Underground, Space Technology, 23(2): 185-198.

Nomoto T, Imamura S, Hagiwara T, et al. 1999. Shield tunnel construction in centrifuge. Journal of Geotechnical and Geoenvironmental Engineering, 125(4): 289-300.

Pastor M L, Balandraud X, Grédiac M, et al. 2008. Applying infrared thermography to study the heating of 2024-T3 aluminum specimens under fatigue loading. Infrar. Phys. &. Technol. , 51: 505-515.

Qiang Z, Xu X, Ning C. 1990. Abnormal infrared thermal of satellite: Forewarning of earthquake. Chinese Sci. Bulletin, 35(17): 1324-1327.

Read R S. 2004. 20 years of excavation response studies at AECL's Underground Research Laboratory. International Journal of Rock Mechanics and Mining Sciences, 41: 1251-1275.

Sagong M, Bobet A. 2002. Coalescence of multiple flaws in a rock-model material in uniaxial compression. International Journal of Rock Mechanics and Mining Sciences,39: 229-241.

Sharma J S, Bolton M D, Boyle R E. 2001. A new technique for simulation of tunnel excavation in a centrifuge. Geotech. Test. J. , 24(4): 343-349.

Shi W Z, Wu Y H, Wu L X. 2007. Quantitative analysis of the projectile impact on rock using infrared thermography. Int. J. Impact Eng. , 34: 990-1002.

Shin J H, Choi Y K, Kwon O Y, et al. 2008. Model testing for pipe-reinforced tunnel heading in a granular soil. Tunnelling and Underground Space Technology, 23(3): 241-250.

Steinberger R, Leitão T I V, Ladstätter E, et al. 2006. Infrared thermographic techniques for non-destructive damage characterization of carbon fiber reinforced polymers during tensile fatigue testing. Int. J. Fatig. , 28: 1340-1347.

Strouth A, Eberhardt E. 2005. The use of LiDER to overcome rock slope hazard data collection challenges at Afternoon Creek, Washington. In: 41st US Symposium on Rock Mechanics, Golden, Colordo. American Rock Mechanics Association, Cdrom.

Tsesarsky M. 2012. Deformation mechanisms and stability analysis of undermined sedimentary rocks in the shallow subsurface. Engineering Geology, 133-134: 16-29.

Tsesarsky M, Hatzor Y H. 2006. Tunnel roof deflection in blocky rock masses as a function of joint spacing and friction-A parametric study using discontinuous deformation analysis (DDA). Tunnelling and Underground Space Technology, 21(1): 29-45.

Wu L X, Liu S J, Wu Y H, et al. 2002. Technical Note: Changes in infrared radiation with rock deformation. International Journal of Rock Mechanics and Mining Sciences, 39: 825-831.

Wu L X, Liu S J, Wu Y H, et al. 2006. Precursors for rock fracturing and failure-Part I, IRR image abnormalities. International Journal of Rock Mechanics and Mining Sciences, 43: 473-482.

Wu L X, Wang J Z. 1998. Technical note: Infrared radiation features of coal and rocks under loading. International Journal of Rock Mechanics and Mining Sciences, 35(7): 969-976.

Wu L X, Wu Y H, Liu S J, et al. 2004. Technical Note: Infrared radiation of rock impacted at low velocity. International Journal of Rock Mechanics and Mining Sciences, 41: 321-327.

Zhao J, Gong Q M, Eisensten Z. 2007. Tunnelling through a frequently changing and mixed ground: a case history in Singapore. Tunnelling and Underground Space Technology, 22(4): 388-400.

Zhi Y, Cui C, Zhang J. 1996. Application of infrared imaging system to the basic remote sensing experiment on rock mechanics. Remote Sens. Environ. Chin. , 11(3): 161-167.

Zhu W S, Li Y, Li S C, et al. 2011. Quasi-three-dimensional physical model tests on a cavern complex under high in-situ stresses. International Journal of Rock Mechanics and Mining Sciences, 48: 199-209.

Zhu W S, Zhang Q B, Zhu H H, et al. 2010. Large-scale geomelchanical model testing of an underground cavern group in a true three-dimensional (3-D) stress state. Canada Geotech J. , 47: 935-946.

Zuo J P, Peng S P, Li Y J, et al. 2009. Investigation of karst collapse based on 3-D seismic technique and DDA method at Xieqiao coal mine. Chin. Int. J. Coal Geol. , 78: 276-287.

Chapter 2   Theoretical aspects of the infrared

Infrared physics, as fundamentals for infrared thermography and its applications in various disciplines, includes many aspects, such as radiometry, photometry, the basic law for thermal radiation, source of the infrared radiation, transmission of the radiation in the atmosphere, infrared measuring and imaging techniques and instruments, etc. For better understanding the principles of the "infrared thermography for geomechanics", the related theoretical aspects are given in Chapter 2 based on the standard publications by Gaussorgues (1994), Maldague (1993), Astarita and Carlomagno (2013), Halliday et al. (2005), Hudson (1969), and Zhang and Fang (2004) etc.

2.1   The infrared

Matter continuously emits and absorbs electromagnetic radiation. The process of emission involves molecular excitations in the material, which generates radiative transitions in its constituent particles. According to Maxwell law of induction, if an electric charge q, which generates the electric field E, is accelerated, the energy will be liberated in the form of radiation (Halliday et al., 2005). A rise in temperature causes an increase in molecular excitation within the material, which favours the acceleration of particular electrical charge carriers and, hence the generation of *electromagnetic radiation* (Gaussorgues, 1994).

Heat transfers by radiation (or *radiative heat transfer*) is an energy transport mechanism that occurs by means of electromagnetic waves. Atoms and molecules constituting a body contain charged particles (protons and electrons) and their movement results in the emission of electromagnetic radiation, which carries energy away from the body surface. Contrary to the case of heat conduction (and consequently convection), energy can be transmitted by thermal radiation also in the absence of a medium and, between two unconnected bodies placed in a vacuum.

The electromagnetic radiation is also referred to as *electromagnetic wave*, because it propagates in the form of waves. The electromagnetic waves travel in the vacuum at a velocity c_0 (3.0×10^8 m/s, independent of λ) and, follows the same laws of reflection, refraction, interference, diffraction, and polarization. The amount of thermal radiation which is absorbed or emitted, as well as its propagation, depends not only on the nature of the material and surface finish but also on its thermodynamic state and on the specific wavelength of the considered electromagnetic wave. All materials at a temperature above absolute zero emit energy by means of electromagnetic waves. At the same time, all materials also absorb electromagnetic waves. Both emission and absorption behaviours are possible because materials change their internal energy state at a molecular level.

The wavelength λ (m) is linked to the frequency of the wave ν (Hz) by the wave speed of propagation (speed of light or electromagnetic wave) c (m/s) in the material (generic medium),

$$c = \nu \lambda \tag{2.1}$$

The speed of propagation in a generic medium is related to the propagation speed in vacuum c_0 by the relationship:

$$c = \frac{c_0}{n} \tag{2.2}$$

where, n is the dimensionless *index of refraction* (or *refractive index*) of the medium, which generally depends also on the wavelength; While both c and λ depend on the nature of the medium through which the wave travels and its thermodynamic state; ν is a constant dependent only on the source of the electromagnetic wave.

The electromagnetic radiation possesses the wave-particle two-phase nature. It obeys the laws of wave propagation and, at the same time, the electromagnetic wave should be viewed as a collection of discrete particles termed *photons* or *quanta* under which the quantum theory applies. Therefore, each photon is considered to have an energy e (J) given by

$$e = h\nu = \frac{hc}{\lambda} \tag{2.3}$$

where, $h = 6.63 \times 10^{-34}$ J, is named *Planck's constant*; From the previous equation, it is clear that, while both c and λ depend on the medium through which the wave travels; ν is constant because the energy of the photon must be conserved.

The energy density, u, of the sinusoidal electromagnetic waves can be written as (Halliday et al., 2005)

$$u = \frac{1}{2}\varepsilon_0 E^2 = \frac{1}{2}\varepsilon_0 E_m^2 \sin(2\pi kx - \omega t) \qquad (2.4)$$

where, E is the electric field; and E_m the amplitude of the field; ε_0 is called permittivity constant, and takes the value of 8.85×10^{-12} C^2/(N · m^2); t is the time variable; ω is the angular frequency; and k is the wave number, given by

$$k = \frac{1}{\lambda} \qquad (2.5)$$

The SI unit for k is the inverse meter. It can be seen that the wavelength of emitted radiation varies inversely with the transition energy. It is important to note that, in the infrared region, the wavelengths are long and the radiation energy is low. Thus difficulties will be encountered in the detection of infrared radiation.

If a medium is present in between the two exchanging bodies, the transferred energy may be partially or completely absorbed and/or reflected, or may even pass through the medium without downgrading. In the latter case, the medium is called *fully transparent* and this practically enables an IR scanner to *view* the temperature of a body without touching it. A medium can also be *partially transparent*, i.e. if it allows only a fraction of the transmitted energy to pass through. However, in many instances, solids and liquids are *opaque* (i.e. completely non-transparent) and, in such cases, the incident non-reflected radiation is absorbed within a few micrometers of their skin. Furthermore, the surrounding molecules absorb the radiation generated within opaque bodies and for these reasons, radiation can be considered as just a *surface* phenomenon.

Therefore, by coating the surface of a body with a very thin layer of opaque material (such as a dull enamel), it is possible to completely change its surface radiation properties and this effect may be very useful when using an IR scanner to measure surface temperatures, particularly of metals (Astarita and Carlomagno, 2013). Clearly, what we affirmed above is true for the wavelengths of interest in the thermal radiation band but it is of course erroneous if one considers for example the X-ray band for which an enamel layer is practically transparent.

2.2　Infrared spectral band

The entire *electromagnetic spectrum* is quite roughly divided into a number of wavelength intervals, called *spectral bands* or more simply *band*, and extends from very small wavelength values ($\lambda \to 0$) to extremely large ones ($\lambda \to \infty$). On inspection of the relevant portion of the electromagnetic spectrum shown in Figure 2.1, the *thermal radiation* band is conventionally defined as a relatively small fraction of the complete spectrum, positioned between 0.1 μm and 1000 μm, which includes part of the *ultraviolet* and all of the *visible* and *infrared bands*.

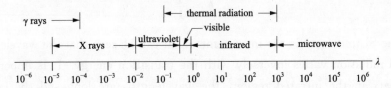

Figure 2.1 Electromagnetic spectrum, the wavelength λ in micrometers (Astarita and Carlomagno, 2013)

It is seen from Figure 2.1 that the infrared radiation only occupies a small fraction of the complete electromagnetic spectrum, starting from the visible red light, extending to the boundary of the microwave. The spectrum of the infrared radiation falls in between 0.75-1000 μm which is a rather wide region composed by frequency sub-regions around 20 order of magnitude. On the contrary, the scope of the visible light is 0.35-0.75 μm comprising of only one octave. When a body is at ambient temperature, most of the energy is radiated in the infrared spectral band. This band is generally sub-divided into four smaller bands with arbitrarily chosen boundaries (see Figure 2.2): *near infrared* (0.75-3 μm), *middle infrared* (3-6 μm), *far* or *long infrared* (6-15 μm) and *extreme infrared* (15-1000 μm). However, it should be noted that not only the boundaries but also the involved semantics might change according to the particular context.

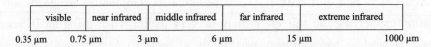

Figure 2.2 Sub-regions of the infrared spectrum

As already stated, if the temperature of a solid, liquid or gas is above absolute zero, their electrons, atoms and molecules are in continuous motion, thus radiation is constantly emitted, absorbed and transmitted through them. Therefore, it has to be stressed that radiation is a *volumetric* phenomenon. The division of the infrared spectrum is under the consideration of the atmospheric transmittance characteristics.

The atmosphere between the source of radiation and the detector is usually the cause of perturbation in infrared detection. The emitted infrared energy is attenuated by the atmosphere, whereas temperature gradients and turbulence create inhomogeneities in the refractive index of air, all of which tend to degrade image quality. Finally, the atmosphere is itself a source of radiation.

The phenomenon of attenuation is a particular problem in the course of measurement, because it introduces a systematic error that depends on the working wavelength, the spectral band used, the distance and weather conditions. The transmission of optical radiation by the atmosphere depends mainly on two phenomena, namely, self-absorption by the atmospheric gases and absorption due to scattering by particles in the air, by molecules and by aerosols. In the former process, the gases absorb radiation selectively at the local temperature. In the latter process, the radiation is deflected from its path and is "seen" by the detector as if it were surrounded by a luminescent veil whose effect depends on the spectral interval used (Gaussorgues, 1994).

Figure 2.3 shows a real slab of finite thickness (partially transparent). Different from a black body (will be addressed in the section 2.5), only a fraction of the incident radiation (often called *irradiation*) is absorbed by the slab. The remaining fraction of the irradiation may be partially reflected and/or partially transmitted across the slab medium. By denoting with a the fraction of irradiation absorbed by the slab, with R the fraction of irradiation reflected and with τ the fraction of irradiation transmitted through the slab, energy conservation requires (Astarita and Carlomagno, 2013),

$$a + r + \tau = 1 \qquad (2.6)$$

where, a, r and τ are respectively called *absorptivity* (or *absorptance*), *reflectivity* (or *reflectance*) and *transmissivity* (or *transmittance*) coefficients (all dimensionless) of the body under consideration.

The transmittance can be explained by the physical model shown in Figure 2.4 Monochromatic radiation is found to be absorbed exponentially by the gas molecules. This is the *Bouguer-Lambert law* which can be verified empirically and can be derived as follows. Consider an absorbing medium and incident radiant flux F_0. A slice of the medium situated at a distance x from the source absorbs an amount dF of the flux, which depends on the thickness dx of the absorbing slice, on the flux F transmitted by it and on the *spectral absorption coefficient* ξ that represents the

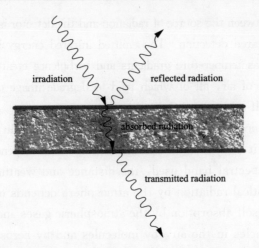

Figure 2.3 Reflection, absorption and transmission of the irradiation in a slab (Astarita and Carlomagno, 2013)

properties of the absorbing medium (Astarita and Carlomagno, 2013),

$$dF = -\zeta F \, dx$$

Integration then yields the attenuation law

$$F = C e^{-\zeta x} \tag{2.7}$$

The constant of integration C is obtained by putting $F=F_0$ at $x=0$,

$$F = F_0 e^{-\zeta x} \tag{2.8}$$

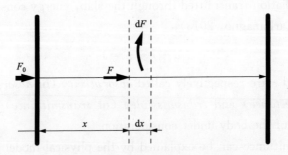

Figure 2.4 Atmospheric absorption

Based on Eq. (2.8), the *transmittance*, τ, of the medium is then defined by,

$$\tau = \frac{F}{F_0} = e^{-\zeta x} \tag{2.9}$$

Chapter 2 Theoretical aspects of the infrared

where, ξ does not depend on the layer thickness but on the wavelength of the incident radiation and on the nature of the material. In general, the coefficients a, r and τ depend on the nature of the material, the surface finish of the object, its thermodynamic state as well as on the wavelength and direction of the impinging radiation.

Figure 2.5 shows the *spectral transmissivity (spectral transmittance)* spectrum (i.e. *transmittance* vs. wavelength denoted by $\tau(\lambda)$ of the atmosphere for the given distance under well-defined weather conditions. The transmission spectrum is shown in the "visible" range up to 13.5 μm. actually, radiation normally called optical extends from 0.18 μm to an upper limit that, for practical purposes, tends to be close to 1000 μm. Radiation that has traversed large expanses of the atmosphere exhibits absorption bands due to water vapour. This constituent of the atmosphere is responsible for infrared absorption; absorption due to the other gases, mostly carbon dioxide (CO_2), is of lower intensity (Gaussorgues, 1994).

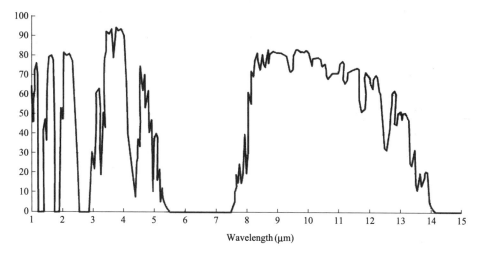

Figure 2.5 Sepctral transmittance of a 1 km thick atmospheric layer in the near, middle and long infrared bands as a function of λ; data from Gebbie et al. (1951)(Astarita and Carlomagno, 2013)

The main absorption bands (see Figure 2.5) due to water vapour lie at about 2.6 μm, between 5.5 and 7.5 μm and beyond 20 μm. They give rise to almost total absorption of radiation over a path length of less than 100m. For practical applications of the infrared thermography, there is a number of *transmission win-*

dows (*atmospheric windows*), i. e., those spectral ranges in which absorption is very weak. These "windows" are: 0.35-1, 1.2-1.3, 1.5-1.8, 2.1-2.5, 3-5, and 7.5-13.5 μm. The significant absorptions (especially in the 5-7.5 μm band) are essentially linked to the presence of water vapour and carbon dioxide. Thus the band between 5 μm and 7.5 μm is seldom used because of its rather high atmospheric absorption (low transmittance).

For thermographic measurements, one has to use the two so-called *atmospheric windows*, which are respectively located between the visible band and about 5 μm, and between 7.5 μm and 13.5 μm spectral bands. They justify the adopted *middle infrared* and *long infrared* bands, the former being downwards limited by the low emissive power at usual temperatures. By considering that when performing laboratory geomechanical measurements the typical measuring distance (i. e., thickness of the air layer) is of the order of a meter, the spectral transmissivity coefficient, τ_λ, is normally quite high, i. e., it can be considered very close to unity.

Most of the currently used IR camera temperature detectors are sensitive in either the middle infrared or the long infrared spectral bands. Detectors are also available in the near infrared band, sometimes sub-divided into very near infrared (0.76-1.1 μm) and short wavelength (1.1-3 μm) bands, and in the extreme infrared band, but they are much less for standard geomechanical applications so these particular bands will not be addressed in the following text.

2.3 Radiometry fundamentals

Radiometry is a subject dealing with the theory and techniques for measurement of the electromagnetic radiant energy, and the basis for thermal emission by matter. The radiometry is built on the basis of optics under the two hypotheses: ①the radiation propagates along a straight line and, therefore, distribution of the radiant energy can not deviate from the optical path prescribed by the geometrical optics;②the radiant energy is not coherent and, thus the interference effect is not considered in the radiometry. The radiation quantities are the objective quantity obtained from the physical measurement, therefore, the concepts of the radiometry applies over the complete electromagnetic spectrum. In the following, only the infrared related concepts of radiometry are introduced (Zhang and Fang, 2004).

2.3.1 Radiant energy

Radiant energy refers to the energy transmitted or received in the form of electromagnetic waves. The radiant energy is denoted by Q, in Joule (J). The radiant energy per unit volume, w in J/m², in the radiation field is defined by,

$$w = \frac{\partial Q}{\partial V} \tag{2.10}$$

where, V is the volume, in m².

Because the radiant energy, w, is also a function of many factors such as the wavelength, area and solid angle, thus the relation of w with respect to Q is given by the partial derivative. For the same reason, the partial derivative is also used to defined the else radiation quantities.

2.3.2 Radiant power and flux

The *radiant power* is the rate of emitted, transmitted and received radiant energy, denoted by P. The unit of the radiant power is watt (abbreviated W). The radiant power is defined by,

$$P = \frac{\partial Q}{\partial t} \tag{2.11}$$

where, t (s) is the time variable.

The *radiant flux* (or *flux*), F, is the instantaneous measure of the quantity of radiation. It describes the output of a source propagating in the form of a beam or received by a detector. The flux manifests itself as radiant power and obeys the laws of propagation in homogeneous non-absorbing media. The unit of the flux, F, is also W. By definition, the physical meaning of the flux is the radiant energy passing through an area per unit time, the same as that of the radiant power.

2.3.3 Geometrical spreading of a beam

Consider a source S and a detector R a distance d apart, and surface elements dS and dR on the source and detector, respectively. Let $d\Omega_S$ be the solid angle subtended by dR at dS and, $d\Omega_R$ the solid angle subtended by dS at dR (see Figure 2.6). We then have (Gaussorgues, 1994)

$$d\Omega_S = \frac{dR\cos\theta_R}{d^2}$$

$$d\Omega_R = \frac{dR\cos\theta_S}{d^2}$$

where, θ_S and θ_R are the angles between the line joining dS to dR and the normals N_S and N_R to dS and dR, respectively. The solid angles are measured in steradians (sr). A complete sphere subtends a solid angle of 4π steradians at its center.

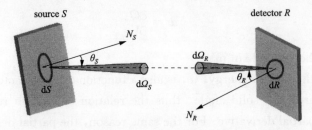

Figure 2.6 Beam geometry (Gaussorgues, 1994)

The geometrical spread of a thin beam subtended by dS and dR is defined by,

$$d^2G = dSd\Omega_S\cos\theta_S = dRd\Omega_R\cos\theta_R = \frac{dSdR\cos\theta_S\cos\theta_R}{d^2} \qquad (2.12)$$

The surface elements dS and dR must be situated in the same optical space (object, image or intermediate). If the surface S is small and R is circular and subtends an angle of 2α at S (see Figure 2.7), the geometrical spread is given by,

$$G = \pi S \sin^2\alpha \qquad (2.13)$$

where, the unit of G is $m^2 \cdot sr$. This form applies to the majority of measuring instruments with circular pupils.

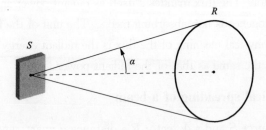

Figure 2.7 Geometrical spread of a beam (Gaussorgues, 1994)

2.3.4 Radiance

If in the neighbourhood of the direction of propagation dS → dR, the flux density is uniform, the *radiance* of a thin beam constrained by dS and dR is defined by (Gaussorgues, 1994),

$$L = \frac{d^2 F}{d^2 G} \tag{2.14}$$

where, $d^2 F$ is the element of flux carried by a thin beam with geometrical spread $d^2 G$. If the medium is homogeneous an non-absorbing, the flux F, the radiance L and the geometrical spread are conserved. The unit of L is $Wm^{-2} \cdot sr^{-1}$.

For an extended beam, we have to integrate these relations over the surfaces of the source and receiver. The integration of the geometrical spread is

$$G = \iint_{SR} d^2 G \tag{2.15}$$

The integration of the radiant flux is

$$F = \iint_{SR} d^2 F = \iint_{SR} L d^2 G \tag{2.16}$$

If the radiance is uniform, then the flux becomes,

$$F = LG \tag{2.17}$$

When the medium in which the radiation propagates is absorbing, the ratio of received to emitted fluxes gives the *transmission factor* (also referred to as *transmissivity* or *transmittance*) of the medium,

$$\tau_r = \frac{d^2 F_R}{d^2 F_S} = \frac{L_R}{L_S} \tag{2.18}$$

The transmittance τ is defined previously in section 2.2 in Eq. (2.9) with the consistent meaning. When $\tau < 1$, corresponding to the absorbing media like the significant absorption of the water vapour and carbon dioxide in the atmosphere over the 5-7.5 μm band; $\tau = 1$ and $\tau > 1$, corresponding to the non-absorbing and emitting media respectively.

For an extended beam, the transmittance is given by,

$$\tau = \frac{F_R}{F_S} = \frac{\iint_{SR} \tau L_S d^2 G}{\iint_{SR} L_S d^2 G} \tag{2.19}$$

The radiance is conserved on refraction,

$$n^2 d^2 G = \text{const}$$

where, n is the refractive index of the medium in which the geometrical spread $d^2 G$ is evaluated. This invariance implies

$$\frac{L'}{n'^2} = \tau \frac{L}{n^2}$$

2.3.5 Irradiance

The *irradiance* is defined as the local value of the ratio of the flux dF_R received by the detector and the area dR of the detector, i. e., the power received per unit area (Wm^{-2}). The irradiance, by its definition as shown in Figure 2.8, can be written as,

$$E(X,Y) = \frac{dF_R}{dR} = \int L(\xi,\eta) \cos\theta_R d\Omega_R \qquad (2.20)$$

where, the integral is evaluated over the half-space, $E(X,Y)$ is in Wm^{-2} and X, Y are the spatial coordinates relative to the points at which the irradiance is evaluated. The quantities ξ, η define the direction of the beam.

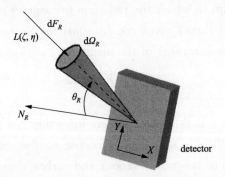

Figure 2.8 Irradiance

2.3.6 Radiant exitance

The *radiant exitance* is defined as the local value of the ratio of the flux emitted by a source and the area of the source. As shown in Figure 2.9, it can be written as,

$$R(X,Y) = \frac{dF_S}{dS} = \int L(\xi,\eta) \cos\theta_S d\Omega_S \qquad (2.21)$$

where, the integral is evaluated over the half-space and $R(X, Y)$ is in Wm^{-2}.

If the source has uniform luminance,

$$R(X,Y) = L \int \cos\theta_S \, d\Omega_S \qquad (2.22)$$

where, the integral is evaluated over the half-space, we have,

$$R = \pi L \qquad (2.23)$$

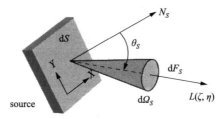

Figure 2.9 Radiant exitance

2.3.7 Radiant intensity of a source in a given direction

This is the ratio of the flux dF_S emitted by a source, in the direction defined the parameters (ξ, η) and the solid angle $d\Omega_S$ within which the intensity is evaluated,

$$I(\xi,\eta) = \frac{dF_S}{d\Omega_S} = \int_{\text{source}} L(X,Y) \cos\theta_S \, dS \qquad (2.24)$$

where, $L(X,Y)$ is the luminance distribution over the source.

The source intensity is therefore the power radiated per unit solid angle.

2.3.8 Bouguer's law

Bouguer's law (see Figure 2.10) is the relation between the irradiance E of a reveiving surface, due to a source S, and the intensity I of that source in the direction of the receiver lying at a distance d,

$$E = \frac{I \cos\theta_R}{d^2} \qquad (2.25)$$

This inverse square dependence on distance from the source is valid if the linear dimensions of the source are small compared with the distance d.

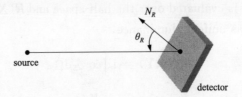

Figure 2.10 Bouguer's law

2.3.9 Radiation scattering

Let us consider an area element dS that receives an element of flux $dF_i(\theta_i, \phi_i)$ in an angular direction defined by (θ_i, ϕ_i) as shown in Figure 2.11. the *reflection coefficient* (*reflectivity* or *reflectance*) of the element dS is given by,

$$\mathcal{R} = \frac{dF_r(\theta_r, \phi_r)}{dF_i(\theta_i, \phi_i)} = \frac{R_r(\theta_r, \phi_r)dS}{E_i(\theta_i, \phi_i)dS} = \frac{R_r}{E_i} \quad (2.26)$$

where, $dF_r(\theta_r, \phi_r)$ is the flux reflected in the direction (θ_r, ϕ_r); and R_r is the exitance of the surface dS under irradiance E_i when reflection is independent of the direction of reemission (orthotropic scatterer). We then have,

$$\mathcal{R} = \frac{R_r}{E_i} \frac{\int L_r(\theta_r, \phi_r)\cos\theta_r d\Omega_r}{E_i} \quad (2.27)$$

where, the integral is evaluated over the half-space, i. e. ,

$$\mathcal{R} = \frac{L_r \int_0^{2\pi} \int_0^{\pi/2} \cos\theta_r \sin\theta_r d\theta_r d\phi_r}{E_i} \quad (2.28)$$

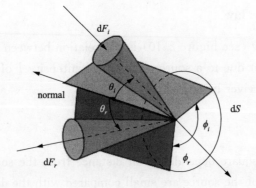

Figure 2.11 Scattering of radiation by a surface

so that,

$$\mathcal{R} = \frac{\pi L}{E} \quad (2.29)$$

In the case of a perfect scatterer, $\mathcal{R}=1$ and $L=E/\pi=R/\pi$, so that

$$E = R \quad (2.30)$$

2.4 Black body radiation

2.4.1 Concept of black body

An opaque body at a specified positive absolute temperature generally emits from its surface thermal radiation in many directions and in a wide range of wavelengths. For a certain body surface temperature, the energy emitted by radiation per unit surface area and unit time (i.e., *energy flux*) depends on the material nature of the body and its surface characteristics, including the surface finish. Clearly, the same reasoning is also true for the absorbed energy.

In order to formulate simple general laws for thermal radiation, it is useful to introduce a conceptual body, usually called a *black body*, which has the property of being a perfect emitter and absorber of radiation. A black body is thus able to absorb all the incident radiation, regardless of its wavelength and direction, and is the body that, for a fixed temperature and wavelength, emits the maximum possible amount of radiation. A black body behaves also as *diffuse emitter*, in the sense that it emits radiation uniformly in all possible directions.

If one considers only the visible part of the spectrum, a black body (at about ambient temperature) can be approximated by a dull black surface (see Figure 2.12) because it absorbs almost the entire incident light without any significant reflection. Since the visible band is a very small part of the complete spectrum, in general a black-coloured surface is seldom a good approximation to an ideal black body.

Instead, a good approximation to a black body is an *isothermal cavity* with a very small aperture, as shown in Figure 2.13 In such a cavity, thermal radiation entering the cavity via the aperture undergoes several reflections before leaving the cavity once again through the aperture. Upon each reflection, part of the incident radiation is absorbed by the surface, therefore, the radiation eventually leaving through the aperture is degraded to an extremely small value. Furthermore, also the radiation emitted by the interior surface of the cavity undergoes many reflections before exiting from the aperture (one can simply reverse the direction of the arrows in Figure 2.13), thus producing a maximum emission in all directions.

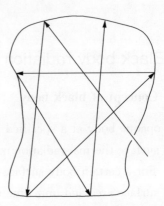

Figure 2.12 A dull black surface element of the radiator

Figure 2.13 Isothermal cavity with a small aperture approximating a black body

2.4.2 Planck's law

The emission of radiation by a black body is described by Planck's law which employs the concepts of statistical thermodynamics and takes the form,

$$R_{b\lambda}(\lambda, T) = \frac{2\pi hc^2 \lambda^{-5}}{\exp[hc_0/(\lambda K_B T)] - 1} \tag{2.31}$$

where, $R_{b\lambda}(\lambda, T)$ is the *spectral exitance*, i.e., the power emitted per unit area per unit wavelength by a black body in vacuum; $h = 6.6256 \times 10^{-34}$ Js (or Ws^2), is Planck's constant; $K_B = 1.38054 \times 10^{-23}$ J/K, is Boltzmann's constant, $c_0 = 3.0 \times 10^8$ m/s, is the speed of light; and T is the absolute temperature of the black body in degrees Kelvin (K).

The temperature conversions are,

$$\text{degrees Celsius} = (°F - 32) \times \frac{5}{9}$$

$$\text{degrees Kelvin} = °C + 273.16$$

$$\text{degrees Fahrenheit} = °C \times \frac{9}{5} + 32$$

Planck's law of radiation can be easily extended to a black body that is emitting in a generic medium by substituting in the expression of each radiation constant the speed of light in vacuum with that in the considered medium [see Eq. (2.2)].

Figure 2.14 shows variation of the spectral exitance of the black body against wavelengths in the temperature range 500-900 K. In the figure, the dashed line indicates the location where the $R_{b\lambda}(\lambda, T)$ attains its maximum value. It is evident that, for each wavelength, the emitted radiation increases significantly with temperature and the different curves never cross each other. Furthermore, the emitted radiation is a continuous function of the wavelength and each isothermal curve tends to zero for both very large and very small values of λ, having a maximum at some intermediate wavelength. Upon an increase in the black body's temperature, the position of this maximum shifts towards small wavelength.

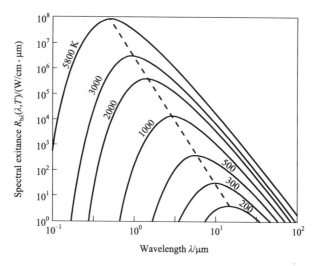

Figure 2.14 Spectral hemispherical black body emissive power (*spectral exitance*), in W/(m² · μm), in vacuum for several absolute temperature (K) values a a function of the wavelength λ (Astarita and Carlomagno, 2013)

It is interesting to note that only at very high temperatures does a significant part of the emitted radiation fall in the visible part of the electromagnetic spectrum (about 0.4-0.76 μm). In fact, as early as in 1847, Draper determined that a thin strip of platinum (but also other materials showed a similar behaviour) could be distinguished, in a dark chamber by the human eye, only when its temperature was over about 800 K. At temperature, the colour of the metal is red because a small part of the emitted energy falls at the very right side of the visible spectrum (Astarita and Carlomagno, 2013). In Figure 2.15, the bands (*middle infrared* and *long infrared*) captured by the infrared detectors of the most commonly used IR cameras are also indicated with the dashed areas.

In the following two extreme cases, the Planck's law [Eq. (2.31)] will takes the different forms.

(1) For the cases of short wavelength or low temperature, i.e., $K_B T \ll hc_0/\lambda$, we can use the approximation,

$$\exp[hc/(\lambda K_B T)] - 1 \propto \exp[hc/(\lambda K_B T)]$$

so that,

$$R_{b\lambda}(\lambda, T) \approx 2\pi hc^2 \lambda^{-5} \exp[-hc/(\lambda K_B T)] \qquad (2.32)$$

This is Wien's law, valid for $\lambda T < 5000$ μm · K.

(2) For the cases of long wavelengths or high temperature, i.e., $K_B T \gg hc_0/\lambda$, the exponential term is small, so that,

$$\exp[hc/(\lambda k T)] - 1 \approx \left(1 + \frac{hc}{\lambda k T} + \cdots\right) - 1 \approx \frac{hc}{\lambda k T}$$

Hence,

$$R_{b\lambda}(\lambda, T) \approx 2\pi ck T \lambda^{-4} \qquad (2.33)$$

This is the Rayleigh-Jeans relation valid for $\lambda T > 10^5$ μm · K but in significant disagreement in the ultraviolet, where the spectral exitance tends to infinity.

It is feasible to simplify Planck's law of radiation by scaling the emissive power with the fifth power of the temperature (in $Wm^{-3} \cdot K^{-5}$),

$$\frac{R_{b\lambda}(\lambda, T)}{T^5} = \frac{c_1}{(\lambda T)^5 (e^{\frac{c_2}{\lambda T}} - 1)} \qquad (2.34)$$

where, c_1 is the *first radiant constant*, in $W \cdot \mu m^4/m^2$,

$$c_1 = 2\pi hc^2 = (3.7415 \pm 0.0003) \times 10^8$$

and c_2 is the *second radiant constant*, in μm · K,

$$c_2 = hc/K_B = (1.43879 \pm 0.00019) \times 10^4$$

According to above relationship, the scaled emissive power (*normalized spectral exitance*) is a function of the sole variable (λT) and, therefore, it can be easily plotted as a single curve (see Figure 2.15, where its normalized value is represented with a solid line). Within this description, it is evident that the curve has a single maximum, for a given λT value. By looking at this curve, or by differentiating Eq. (2.31) with respect to λ, or differentiating Eq. (2.34) with respect to λT and setting to zero the result, Wien's displacement law can be obtained (see the following text).

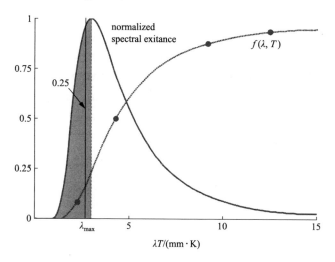

Figure 2.15 Normalized spectral hemispherical black body emissive power (*normalized spectral exitance*) in vacuum and fractional emissive power function with selected points (after Astarita and Carlomagno, 2013)

2.4.3 Wien's law

Wien's displacement law (also Wien's law) can be obtained by differentiating Planck's law with respect to the wavelength, and then, let the derivative equal to zero,

$$\frac{\mathrm{d}R_{b\lambda}(\lambda,T)}{\mathrm{d}\lambda}=\frac{\mathrm{d}}{\mathrm{d}\lambda}\left(\frac{c_1}{\lambda^5}\frac{1}{\mathrm{e}^{c_2/\lambda T}-1}\right)=0 \tag{2.35}$$

from Eq. (2.34), we can obtain,

$$1-\frac{c_2}{5\lambda_{\max}T}=4.965$$

By using the method of successive approximation, obtained is the expression for Wien's law,

$$\lambda_{\max}=\frac{2898}{T} \tag{2.36}$$

where, T is in degrees Kelvin; and λ_{max} in μm. This equation represents Wien's displacement law which enables one to find the wavelength λ_{max} at which a black body emits its maximum spectral emissive power as a function of its temperature.

The Wien's law demonstrates the fact that the peak value wavelength λ_{max}, at which the black body's spectral exitance attains its maximum value, is inversely proportional to the absolute temperature, T.

For example, an object at an ambient temperature $T \approx 290$ K has its spectral exitance maximum at $\lambda_{max} \approx 10$ μm, while the Sun, whose apparent temperature is 6000 K, emits radiation peaking at $\lambda_{max} = 0.5$ μm (this wavelength is exactly the center of the visible spectrum). Therefore, the more that fifty percent of the Sun's radiation power fall into the visible band and ultraviolet band. Human body ($T=310$ K) emits radiation peaking at $\lambda_{max} \approx 9.4$ μm, almost all the radiant power of the human body falls into the infrared band. A liquid nitrogen ($T=77$ K) produces maximum radiation at $\lambda_{max} = 38$ μm.

2.4.4 Stefan-Boltzmann law

Stefan-Boltzmann law is obtained by integrating of Planck's law over the whole spectrum (between $\lambda=0$ and $\lambda=\infty$). It gives the *total exitance* of a black body at given temperature T,

$$R_b = \int_0^\infty R_{b\lambda}(\lambda, T) d\lambda = \sigma T^4 \tag{2.37}$$

where, R_b is energy flux over all wavelength (in Wm^{-2}) referred to as the *total exitance* (also *total black body hemispherical emissive power*); and

$$\sigma = \pi^4 c_1 / 15 c_2^4 = 5.670 \times 10^{-8}$$

is known as the *Stefan-Boltzmann constant*, , in $W/(m^2 K^4)$. Stefan-Boltzmann law indicates the fact that the *total exitance* of a black body is proportional to the fourth power of its absolute temperature. Thus, a small change of the temperature will induce very large variation of the *total exitance*.

The area below a curve in Figure 2.15 corresponds to the total exitance of the black body represented by the curve. It is seen that the area below each of the curves increase significantly with the increase of the temperature.

2.4.5 Exitance of a black body in a given spectral band

While making measurements with IR thermography, since (as already seen in Figure 2.14) infrared camera detectors capture only a limited band of the whole electromagnetic spectrum, a definite integral of Planck's law in Eq. (2.31) is more appropriate to find the energy flux sensed by the detector in a certain band λ_1-λ_2, i.e., the *exitance of a black body* between λ_1 and λ_2, which is given by

$$R_{b,\lambda_1-\lambda_2}(T) = \int_{\lambda_1}^{\lambda_2} R_{b\lambda}(\lambda, T) d\lambda \tag{2.38}$$

Figure 2.16 shows the physical meaning for Eq. (2.38), Planck's law and Stefan-Boltzmann's law. The curve in black solid line represents the spectral exitance (Planck's law), the area below the curve represents the Stefan-Boltzmann's law and the shadowed region indicates Eq. (2.38), i.e., the black body exitance within λ_1-λ_2 band.

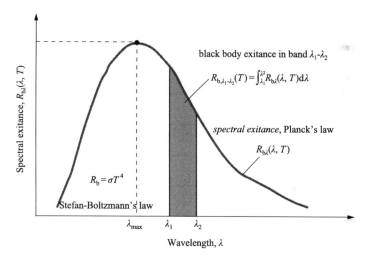

Figure 2.16 Exitance of black body

Table 2.1 gives the values of exitance of a black body, calculated for the spectral band $\Delta\lambda = \lambda_1 - \lambda_2$ using Eq. (2.38) at temperatures normally encountered in thermography.

As aforementioned, the electromagnetic wave can be viewed as a collection of discrete particles termed photons or quanta and the energy carried by a photon is given by Eq. (2.3). In another words, the emission occurs between discrete energy levels by the emitted quantum. i.e., photon. The *spectral energy exitance* given by Planck's law can be expressed in terms of the number of photons emitted per second per square centimeter of radiating surface per unit wavelength, obtained by dividing

Table 2.1 Values of black body exitance for the spectral band $\Delta\lambda=\lambda_1-\lambda_2$
(data is from Gaussorgues, 1994)

$\lambda/\mu m$		Exitance of a black body in $\Delta\lambda$, $R_{b,\lambda_1-\lambda_2}(T)/(W/cm^2)$					
λ_1	λ_2	$T=280$ K	$T=290$ K	$T=300$ K	$T=310$ K	$T=750$ K	$T=1000$ K
3	5	2.76×10^{-4}	4.11×10^{-4}	5.97×10^{-4}	8.48×10^{-4}	5.84×10^{-1}	
3	5	5.44×10^{-4}	7.87×10^{-4}	1.11×10^{-3}	1.54×10^{-3}	7.09×10^{-1}	
3.5	5	2.68×10^{-4}	3.97×10^{-4}	5.75×10^{-4}	8.13×10^{-4}	4.42×10^{-1}	
3.5	5.5	5.36×10^{-4}	7.73×10^{-4}	1.09×10^{-3}	1.50×10^{-3}	5.68×10^{-1}	2.38
4	5	2.38×10^{-4}	3.49×10^{-4}	5.01×10^{-4}	7.02×10^{-4}	2.89×10^{-1}	
4	5.5	5.06×10^{-4}	7.25×10^{-4}	1.02×10^{-3}	1.39×10^{-3}	4.15×10^{-1}	
8	10	4.20×10^{-3}	5.12×10^{-3}	6.15×10^{-3}	7.32×10^{-3}	1.74×10^{-1}	
8	12	8.59×10^{-3}	1.03×10^{-2}	1.22×10^{-2}	1.43×10^{-2}	2.74×10^{-1}	
8	14	1.26×10^{-2}	1.48×10^{-2}	1.74×10^{-2}	2.01×10^{-2}	3.34×10^{-1}	6.05
10	12	4.39×10^{-3}	5.17×10^{-3}	6.02×10^{-3}	6.95×10^{-3}	9.99×10^{-2}	
10	14	8.35×10^{-3}	9.72×10^{-3}	1.12×10^{-2}	1.28×10^{-2}	1.60×10^{-1}	
12	14	3.96×10^{-3}	4.55×10^{-3}	5.19×10^{-3}	5.86×10^{-3}	6.04×10^{-2}	

$$R_{pb\lambda}=R_{b\lambda}(\lambda,T)\frac{\lambda}{hc}=\frac{2\pi c\lambda^{-4}}{\exp[hc/(\lambda kT)]-1} \quad (2.39)$$

where, $R_{pb\lambda}$ is the numbers of photon emitted into hemispherical space unit time unit area, in $s^{-1}\cdot cm^{-2}\cdot K^{-3}$.

Under these conditions, Wien's law becomes,

$$\lambda'_{max}=\frac{3663}{T}(\mu m) \quad (2.40)$$

$$R_{pb\lambda}(\lambda'_{max},T)=2.1\times10^7\times T^4(s^{-1}\cdot cm^{-2}\cdot K^{-3}) \quad (2.41)$$

and Stefan-Boltzmann's law becomes,

$$R_{pb\lambda}=\sigma'T^3 \quad (2.42)$$

where,

$$\sigma'=1.52\times10^{11}(s^{-1}\cdot cm^{-2}\cdot K^{-3})$$

2.4.6 Calculation of exitance of black body

In order to simplify the evaluation of this integral it is convenient to introduce the *dimensionless fractional emissive power function* based on Eq. (2.38),

Chapter 2 Theoretical aspects of the infrared

$$f(\lambda T) = \frac{R_{b,0-\lambda}}{R_{b,0-\infty}} = \frac{\int_0^\lambda R_{b\lambda}(\lambda,T)d\lambda}{\int_0^\infty R_{b\lambda}(\lambda,T)d\lambda} = \int_0^{\lambda T} \frac{R_{b\lambda}(\lambda,T)}{\sigma T^5} d(\lambda T) \quad (2.43)$$

where, the integrand of the last term of the previous equation is a function of the only variable λT. The function defined by Eq. (2.43) is plotted in Figure 2.15 with a dashed line. Particular percentages (10%, 50%, 90% and 95%) of emitted flux are also indicated on this line with the easily recognizable dots.

A closed form of the fractional emissive power function was presented by Chang and Rhee (1984),

$$f(\lambda T) = \frac{15}{\pi^4} \sum_{n=1}^{\infty} \left[\frac{e^{-n\eta}}{n} \left(\eta^3 + \frac{3\eta^2}{n} + \frac{6\eta}{n^2} + \frac{6}{n^3} \right) \right] \quad (2.44)$$

where, $\eta = c_2/\lambda T$. The definite integral defined by Eq. (2.38) and Eq. (2.43) can now be used to evaluated the *exitance of a black body* between λ_1 and λ_2,

$$R_{b,\lambda_1-\lambda_2}(T) = \sigma T^4 [f(\lambda_2 T) - f(\lambda_1 T)] \quad (2.45)$$

The exitance of a black body between 0 and λ,

$$R_{b,0-\lambda}(T) = f(\lambda T) R_{b,0-\infty} = \sigma T^4 f(\lambda T) \quad (2.46)$$

It is interesting that calculating the radiation of some commonly encountered phenomena by using the fundamental laws introduced previously. For example, the human body, under the skin of a human body is black body at temperature $T=310$ K, the peak-value waveleng, λ_{max} at which a black body emits its maximum spectral emissive power as a function of its temperature, is given by the Wien's law,

$$\lambda_{max} = \frac{2898}{T} = \frac{2898}{310} = 9.4(\mu m)$$

The *total exitance* of a black body is given by the Stefan-Boltzmann's law,

$$R = \sigma T^4 = 5.67 \times 10^{-8} \times 310^4 = 5.2 \times 10^2 (Wm^{-2})$$

The exitance in the ultraviolet region between wavelength $\lambda = 0-0.4\ \mu m$ is given by Eq. (2.46),

$$R_{0-0.4} \cong 0$$

The exitance in the visible light region between wavelength $\lambda = 0.4-0.75\ \mu m$ is given by Eq. (2.45),

$$R_{0.4-0.75} \cong 0$$

The exitance in the infrared region between wavelength $\lambda = 0.75$ μm-∞,

$$R_{0.75-\infty} \cong R$$

Another example is the Sun viewed as a black body at $T=6000$ K, its peak wavelength is,

$$\lambda_{max} = \frac{2898}{6000} = 0.48(\mu m)$$

The total *exitance*,

$$R = \sigma T^4 = 5.67 \times 10^{-8} \times 6000^4 = 7.3 \times 10^7 (Wm^{-2})$$

The exitance in the ultraviolet region between wavelength $\lambda = 0$-0.4 μm is,

$$R_{0-0.4} \cong 0.14R$$

The exitance in the visible light region between wavelength $\lambda = 0.4$-0.75 μm is,

$$R_{0.4-0.75} \cong 0.42R$$

The exitance in the infrared region between wavelength $\lambda = 0.75$ μm-∞,

$$R_{0.4-0.75} \cong 0.44R$$

As already affirmed, infrared scanners can operate in different spectral bands and the previous equation can be used to compute the fraction of the total energy by the black body in the band of interest. It is quite interesting to evaluate this fraction for the two bands most typically used by IR scanners for detection in mechanics experiments, i. e. , the 3-5 μm band of the *middle infrared* scanners and the 8-12 μm band of the *long infrared* ones.

For a black body temperature of 300 K, it is found that the energy radiated in the *long infrared* band is about 26% of the total, while for the *middle infrared* band this percentage is reduced to 1.3%. In this case, the long infrared band appears to have clear advantage but, by increasing the temperature to 600 K, the middle infrared band behaves a little better, i. e. , 23% for middle infrared against 21% for long infrared.

By using the previous equation in combination with Wien's law and/or by looking at the curves of Figure 2.15, one can determine that, irrespective of the temperature of the black body, the percentage of emissive power radiated at wavelength smaller than λ_{max} is about 25% of the total. It is noted that a simple calculation shows that three quarters of the total energy emitted by a black body at temperature T is situated in the spectral band between $\lambda_{\max}$ and infinity.

2.4.7 Thermal radiation contrast

Previous discussions about the fundamental laws of infrared radiation are in fact from the view point of the physics. It focuses on the investigation of the magnitude of the emission power and characteristics of the distribution of the the emitted energy as a function of the wavelength. In this sub-section, we discuss the frequently used concept of *thermal radiation contrast* (also *thermal contrast*) from an engineering application point of view.

Detection an object in the background using infrared thermography will be very difficult when the differences between the radiant exitance from the object and background are small. For quantifying the difference between the radiations from the object and background, the *thermal contrast*, as an indicator, is introduced.

The *thermal contrast* is defined as the ratio of the difference of the radiance exitance by the object in view and the background to that of the background,

$$C = \frac{R_b(T_T) - R_b(T_B)}{R_b(T_B)} \tag{2.47}$$

where, $R_b(T_T)$ is the radiant exitance of the *target* between the wavelength region of λ_1 to λ_2, given by,

$$R_b(T_T) = \int_{\lambda_1}^{\lambda_2} R_{b\lambda}(\lambda, T) d\lambda \tag{2.48}$$

and $R_b(T_B)$ is the radiant exitance of the *background* between the wavelength region of λ_1 to λ_2, given by,

$$R_b(T_B) = \int_{\lambda_1}^{\lambda_2} R_{b\lambda}(\lambda, T) d\lambda \tag{2.49}$$

Optimal thermal contrast could be obtained by selecting the appropriate wave bands in the infrared imager system. Let terperature of the target, T_T, be 310 K, temperature of the background, T_B, be 300 K, and the target and background be viewed as the black bodies, the *full waveband thermal contrast* can be calculated by Eq. (2.47),

$$C_{0-\infty} = \frac{R(T_T) - R(T_B)}{R(T_B)} = \frac{\Delta R(T)}{R(T_B)} = \frac{(\partial R/\partial T)\Delta T}{\sigma T_B^4}$$

where, we omit the subscript b for the radiant exitance R. According to the Stefan-Boltzmann's law, i. e., $R = \sigma T^4$, and hence $\partial R/\partial T = 4\sigma T^3$ under the assumption that ΔT is small. Thus we can obtain,

$$C_{0-\infty} = \frac{4\sigma T^3 \Delta T}{\sigma T^4} = \frac{4\Delta T}{T} = \frac{4 \times 10}{300} = 0.133$$

The thermal contrast in the middle infrared 3.5-5 μm, $C_{3.5\text{-}5\mu m}$, and that in the long infrared 8-14 μm, $C_{8\text{-}14\mu m}$, can be can be calculated using Eq. (2.47),

$$C_{3.5\text{-}5\mu m} = 0.413, \text{ and } C_{8\text{-}14\mu m} = 0.159$$

It is clear that the thermal contrast in the 3.5-5 μm band is greater than in the 8-14 μm band.

2.5 Radiation of real bodies

2.5.1 Different types of radiator

The radiation characteristics of real bodies are normally different from those of a black body. Both emitted and absorbed radiations of a black body are upper limits for real bodies that may possibly be approached only in certain spectral bands and under certain conditions.

The incident radiation (often called *irradiation*) is completely absorbed by a black body but, as shown in Figure 2.3, for a real slab of finite thickness, only a fraction of the irradiation is absorbed. The remaining fraction of the irradiation may be partially reflected and/or partially transmitted across the slab medium. As expressed by Eq. (2.6), non-black bodies absorb only a fraction a of incident radiation, they reflect a fraction r and transmit a fraction τ. These different fractors are selective, i. e., they depend on the wavelength.

Consider an object of this kind, exposed to a given amount of incident radiation (see Figure 2.17). When the system is in a state of thermodynamic equilibrium, the energy released into the ambient medium as radiation plus energy reflected and transmitted, must equal the energy introduced into the system by absorption.

It is thus necessary to introduce the *spectral emissivity* $\varepsilon(\lambda)$ whose role is to balance the *absorptance* (or *absorptivity*) $a(\lambda)$. For a *diffuse emission* (or a *diffuse irradiation*), Kirchhoff's law states that the spectral emissivity is equal to the spectral absorptivity coefficient,

Chapter 2 Theoretical aspects of the infrared

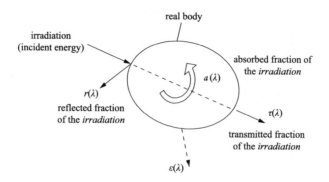

Figure 2.17 Radiant energy balance for non-black bodies

$$a(\lambda) = \varepsilon(\lambda) \tag{2.50}$$

Thus the Eq. (2.6) can be rewritten as,

$$\varepsilon(\lambda) + r(\lambda) + \tau(\lambda) = 1 \tag{2.51}$$

According to the law of variation for the spectral emissivity, $\varepsilon(\lambda)$, following thermal radiators could be quantified, i. e. ,

(1) black body (or Planck's radiator):

$$\varepsilon(\lambda) = 1, \text{ and } r(\lambda) = \tau(\lambda) = 0 \tag{2.52}$$

Black body follows the Planck's equation, Wien's law and Stefan-Boltzmann's law discussed in the previous sections.

(2) grey body: its emissivity independent of the wavelength λ, i. e. ,

$$\varepsilon(\lambda) = \text{const, and } \varepsilon(\lambda) \leqslant 1; \; r(\lambda) = \text{const} \tag{2.53}$$

(3) opaque body: for an opaque body, no incident radiation could be transmitted, and as stated above, for a diffuse emission (or a diffuse irradiation), the spectral emissivity is equal to the spectral absorptivity coefficient (Kirchhoff's law), thus

$$\tau(\lambda) = 0, \text{ and } \varepsilon(\lambda) + r(\lambda) = 1 \tag{2.54}$$

(4) shiny body:

$$r(\lambda) \text{ large, and } \varepsilon(\lambda) \text{ almost zero} \tag{2.55}$$

where, Eq. (2.55) manifests the fact that a shiny body not only emit less energy but also reflect a large amount of the radiation coming from the ambient environment and imaging on them. Example of the shiny bodies is the shiny metallic surfaces having low emissivity. Whenever possible, they should not be employed in infrared applications or, if they must be used for some reason and transient heat transfer is not involved, the viewed surface should be converted with a thin layer of thermally black paint such as dull enamel. Investigation showed that the white enamel has a slightly higher emissivity than black one found in long infrared band (Astarita and Carlomagno, 2013).

(5) selective body: its spectral emissivity is a function of the wavelength λ. Figure 2.18 shows the spectral emissivity curves of typical emitters including the black body, grey body, selective body, shiny body and mirror. It is seen that the spectral emissivity curve of black body is the envelop curve for the else emitters.

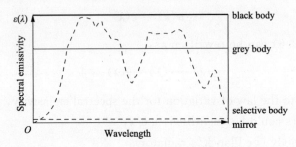

Figure 2.18 Spectral emissivity

From the figure, following phenomena could be observed. The total exitance or spectral exitance of black body in an arbitrary wave band is larger than that of the any other emitters. The emissivity of grey body is a constant, which is a very useful concept. It is because some radiant sources, such as jet-tail nozzle, aerodynamic heating surface, no-powered spacecraft, human body, earth and space background, could be viewed as grey bodies. Their exitance can be calculated using the previous laws if the emissivities are known. The spectral emissivity of the grey body is less than that of the black body, and therefore, its emissivity curve lies below the emissivity curve for the black body. In a finite interval of wavelength, selective body can be viewed as grey body for the simplification of the calculation.

2.5.2 Emissivity of a material

Normal objects are not generally black bodies and the emission from real bodies does not definitely follow Planck's and consequent laws. As already mentioned, the radiation emitted by a black body is an upper limit for real bodies, that may possibly be approached only in certain spectral bands and under certain conditions.

The *emissivity* (or *specific radiance*) is defined as the ratio of the radiation of a body at given temperature T to that of black body at the same temperature. Obviously, large emissivity indicates the fact that the emission of that body approaches black body. If the emissivity of an object is known, the radiation laws or emission parameters of that body can be established using the basic law of radiation for the black body.

On a total basis, a real body generally emits only a fraction R of the heat flux R_b emitted by a black body at the same temperature. This happens to be true also on the spectral basis, i.e., $R_\lambda \leqslant R_{b\lambda}$. In the following, for sake of ease, R is always referred to as the *total hemispherical emissive power* and R_λ is the *spectral hemispherical emissive power*.

The *spectral hemispherical emissivity* ε_λ (dimensionless) is defined by,

$$\varepsilon_\lambda = \frac{R_\lambda(\lambda, T)}{R_{b\lambda}(\lambda, T)} \tag{2.56}$$

where, $R_\lambda(\lambda, T)$ is the spectral exitance of a real body at given temperature T and $R_{b\lambda}(\lambda, T)$ is the spectral exitance of black body at the same temperature. ε_λ can be calculated by substituting Planck's law in Eq. (2.56).

Similarly, the *total hemispherical emissivity*, ε is defined as,

$$\varepsilon = \frac{R(T)}{R_b(T)} \tag{2.57}$$

where, $R(T)$ is the total exitance of a real body at given temperature T; and $R_b(T)$ is the total exitance of a black body at the same temperature. By substituting Stefan-Boltzmann's law into Eq. (2.57), estimation of ε can be made.

The emissivity coefficient is also a function of the angle θ (see sub-section 2.3.3) between the direction normal to the emitting surface and the direction of the emitted radiation. It is referred to as *directional emissivity*, ε_θ, or *spectral directional emissivity*, $\varepsilon_{\theta\lambda}$ (also a function of the wavelength at the same time). When the dependency on the angle θ does not occur, the body is called a *diffuse emitter* (or *Lambert body*).

The black body is a diffuse emitter because its *directional emissivity*, *spectral directional emissivity*, *total hemispherical emissivity* and *spectral hemispherical emissivity* are equal to one, i.e.,

$$\varepsilon_\theta = \varepsilon_{\theta\lambda} = \varepsilon = \varepsilon_\lambda = 1$$

The black body, therefore, is the diffuse emitter and its emissivity is independent of the direction. Else emitters, apart from the polished metal surfaces, are close to the diffuse emitter. The differences between the four types of their emissivity coefficients are very small or even negligible. Therefore, the generic forms of the emissivity are the *emissivity* ε (or total emissivity) or spectral emissivity ε_λ. Values of emissivity for the commonly used materials are reported in Table 2.2.

Table 2.2   Emissivity of commonly used materials

Metal		Temperature/℃	ε	Material		Temperature/℃	ε
steel				concrete		20	0.92
	8%Ni-18%Cr	500	0.35	wood			
	soft	1600-1800	0.28		natural		0.5-0.7
	galvanised	20	0.28		board	20	0.8-0.9
	oxidized	200-600	0.80	red brick		20	0.93
	rusty	20	0.69	rubber			
	polished	100	0.07		hard	20	0.95
	stainless	20-700	0.16-0.45		soft	20	0.86
aluminium				carbon			
	polished	50-500	0.04-0.06		fibre	1000-1400	0.53
	coarse surface	20-50	0.06-0.67		graphite	20	0.98
	alumina	50-100	0.2-0.3		lampblack	20-400	0.95-0.97
	anodized	100	0.55		charcoal		0.96
sliver				lime			0.3-0.4
	polished	200-600	0.02-0.03	cement			0.54
bronze				leather			0.75-0.80
	polished	50	0.10	water			
		500-1000	0.28-0.38		distilled	20	0.96
copper					smooth ice	−10	0.95
	polished	100	0.03		frosted	−10	0.98
	oxidized	50	0.10		snow	−10	0.85
	at melting pt.	1100-1300	0.13-0.15	ebonite			0.89
Tin				enamel			0.9
	polished	20-50	0.04-0.06	tar			0.79-0.84

续表

Metal	Temperature/℃	ε	Material	Temperature/℃	ε
dioxide		0.40	lubricating oil	20	0.82
iron			human skin	32	0.98
coarse unoxidized	20	0.24	paper		
electropolished	200	0.06	white	20	0.7-0.9
rusty	20	0.61-0.85	yellow		0.72
oxidized	100	0.74	red		0.76
galvanized	30	0.25	deep blue		0.84
polished	400-1000	0.14-0.38	green		0.85
cast iron			black		0.9
crude	50	0.81	matt		0.93
liquid	1300	0.28	oil paint		
polished	200	0.21	different colours	100	0.92-0.96
brass			(matt black, 3M, velvet coat)		0.98
polished	100	0.03	plaster	20	0.91
oxidized	200-600	0.60	porcelain	20	0.92 0.7-0.75
magnesium			quartz(fused)	20	0.93
polished	20	0.07	Sand	20	0.6-0.9
powder		0.86	clinker(furnace)	0-100 200-500 600-1200 1400-1800	0.97-0.93 0.89-0.78 0.76-0.7 0.69-0.67
mercury	0-100	0.09-0.12	silica powder		0.48
nickel			soil		
polished	20	0.05	dry	20	0.9
oxidized	200-600	0.37-0.48	wet	20	0.95
gold			talcum powder		0.24
polished	100	0.02	fired clay	70	0.91
platinum			glass(polished)	20 20-100 250-1000 1100-1500	0.94-0.91 0.87-0.72 0.7-0.67

续表

Metal	Temperature/℃	ε	Material	Temperature/℃	ε
polished	200-600 1000-1500	0.05-0.10 0.14-0.18	black cloth	20	0.98
lead			magnesium powder		0.2-0.3
grey oxidized	20	0.28	grey marble (polished)	20	0.93
shiny	250	0.08			
tungsten	20 600-1000 1500-2200 3300	0.05 0.10-0.16 0.24-0.31 0.39	black body (commercial)		0.99
zinc	200-300	0.04-0.05			
polished	400	0.11			
oxidized	1000-1200	0.5-0.6			
powder		0.82			
sheet	50	0.20			

A generic laws of the emissivity variations of the materials could be stated as follows.

(1) For a *diffuse emitter* (or *Lambert body*), the three types of the emissivity, the *normal emissivity* ε_n, *directional emissivity* ε_θ, and *total emissivity* ε, are equal,

$$\varepsilon_n = \varepsilon_\theta = \varepsilon$$

The normal emissivity, ε_n, is referred to the emissivity of a material given at normal incidence.

For dielectric materials, the ratio, $\varepsilon_n/\varepsilon$, falls into the range of 0.95-1.05, averaged 0.98. Thus, when the angle θ is less than 65° or 70°, the directional emissivity and normal emissivity are still equal, i.e., $\varepsilon_\theta = \varepsilon_n$.

For conductive materials, the the ratio, $\varepsilon_n/\varepsilon$, falls into the range of 1.05-1.33. For shiny metallic surfaces, this ratio is averaged 1.20, their hemispherical emissivity is larger than their normal emissivity by 20%, i.e., when the angle is larger than 45°, the difference between the directional emissivity ε_θ and normal emissivity will be significant.

(2) The emissivity of metals is generally low, and increases with temperature. This increase continues until the surface of the warm metal becomes oxidized.

(3) Non-metallic materials have high emissivity values, often above 0.8. They decrease with temperature. In all cases, the emissivity depends on the state of the surface material.

(4) Radiation from the metallic and the else opaque materials occurs within vicinity at several micro-milimeters of the viewed surface. The emissivity, therefore, is a function of the state of the surface and independent of size of the surface. Accordingly, the emissivities for coated or painted surfaces are the properties of the coating and painting layers, having nothing to do with the materials being coated and painted. For a material, the detected emissivity will be vary due to the surface conditions.

2.5.3 Stefan-Boltzmann's law for grey body

Bodies having their emissivity independent of the wavelength λ are called *grey bodies*. Even if no real surface is truly grey over the whole electromagnetic spectrum, often a real surface can have an almost constant spectral emissivity in the used IR detector band so that, at least from a practical detection point of view, the grey hypothesis can be assumed to be satisfied.

All the emissivity and spectral emissivity for a grey body are constant less than 1. By denoting the grey body with subscript g, the laws for black body radiation in section 2.4 can be rewritten for representing the grey body radiation.

The *radiant exitance* of a grey body is given by,

$$R_g(T) = \varepsilon R_b(T) \tag{2.58}$$

The *spectral radiant exitance* of a grey body is,

$$R_{g\lambda}(\lambda, T) = \varepsilon R_{b\lambda}(\lambda, T) \tag{2.59}$$

The *radiance* of a grey body is,

$$L_g = \varepsilon_\theta L_b \tag{2.60}$$

The *spectral radiance* of a grey body is,

$$L_{\lambda g} = \varepsilon_\theta L_{\lambda b} \tag{2.61}$$

When the grey body is a diffuse emitter, as stated previously, the directional emissivity equals to the total emissivity, i.e., $\varepsilon_\lambda = \varepsilon$, then Stefan-Boltzmann's law for grey bodies can be given based on that for black body. That is, the *spectral exitance* for grey bodies is given by,

$$R_{g\lambda}(\lambda, T) = \varepsilon R_{b\lambda} = \frac{\varepsilon c_1}{\lambda^5}\left[e^{c_2/(\lambda T)} - 1\right] \qquad (2.62)$$

The *total exitance* for grey bodies is,

$$R_g(\lambda, T) = \varepsilon R_b = \varepsilon \sigma T^4 \qquad (2.63)$$

The Wien's law expressed in Eq. (2.35) retains its original form for grey bodies.

2.5.4 Dielectric materials

The electromagnetic theory can be used to find the dependence of the spectral directional emissivity on the refractive index n and the dimensionless *extinction coefficient* κ (an imaginary part of the complex refractive index, linked to ξ) of the material. Both n and ξ are a function of the thermodynamic state and electrical properties of the material, in particular of the electrical resistivity r_e as well as of the radiation wavelength.

Dielectrics are materials that conduct electricity poorly include most of the liquids, plastics, paints, glasses, woods, metal oxides, and rock or rock-like materials. As this book is designed to discuss "infrared thermography for geomechanics", discussion of the emission characteristics for dielectric materials is of significance. The radiation behaviour of dielectric materials and that of metals are different from each other. In particular, the electrical resistivity is supposed to be extremely large in the former case (ideal dielectric) while relatively small in the latter. In the following, it is also assumed that the radiation is emitted in air, it being a good approximation to vacuum since normally, for air, $n \approx 1$ with less than a per thousand accuracy (e.g. Gladstone-Dale's law, see Astarita and Carlomagno, 2013).

Normally, for dielectric materials, the extinction coefficient κ is very small and the refractive index n is less than 3. According to the theory, the spectral (being n a function of λ) directional emissivity coefficient $\varepsilon_{\lambda\theta}$, for a smooth opaque medium, can be calculated with the relationship (Baehr and Stephan, 2006),

$$\varepsilon_{\lambda\theta} = \frac{2\cos\theta \sqrt{n^2 - \sin^2\theta}}{(\cos\theta + \sqrt{n^2 - \sin^2\theta})^2}\left[1 + \frac{n^2}{(\cos\theta \sqrt{n^2 - \sin^2\theta} + \sin^2\theta)^2}\right] \qquad (2.64)$$

which, for $n=1$, gives $\varepsilon_{\lambda\theta}=1$, independent of θ.

A plot of the dependence on θ of the emissivity coefficient, as predicted by the previous equation is shown in Figure 2.19 for some values of the refractive index $n>1$. For the considered n values, the emissivity coefficient is almost constant for relatively small θ values (practically up to about 45°) but decrease rapidly when the emitted radiation tends towards being parallel to the surface ($\theta=90°$). With increasing refractive index, the maximum value of $\varepsilon_{\lambda\theta}$ decreases, it later drop being more abrupt and, consequently, confined to a smaller high θ values range.

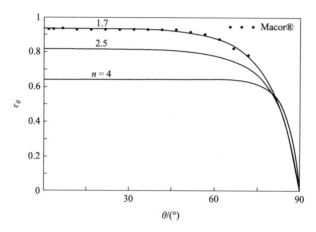

Figure 2.19 Spectral directional emissivity coefficient of dielectric materials for some n values; (after Astarita and Carlomagno, 2013); black dots are experimental results obtained by Dello (2008)

The limit of Eq. (2.64) for $\theta \to 90°$ gives the spectral emissivity coefficient, for a radiation flux *normal* to the emitting surface ($\varepsilon_{\lambda\theta=0}=\varepsilon_{\lambda n}$), a condition that can be relatively easily encountered in experiments:

$$\varepsilon_{\lambda\theta=0} = \varepsilon_{\lambda n} = \frac{4n}{(n+1)^2} = 1 - \left(\frac{n-1}{n+1}\right)^2 = 1 - r_{\lambda n} \qquad (2.65)$$

Obviously, the term in brackets of the third equality member of Eq. (2.65) is the normal spectral reflectivity coefficient $r_{\lambda n}$ (for $\theta \to 0$).

From Eq. (2.65) (but also from Figure 2.20) and since, as already said, generally $n<3$, high normal spectral emissivity values, larger than 0.75, are expected for dielectric materials. This is effectively observed in experimental measurements for $\lambda>2$ μm, i.e., in the bands of the electromagnetic spectrum normally used in most infrared thermography applications. Furthermore, it may be affirmed that, often within each of these bands, dielectrics frequently approximate the behaviour of a grey body.

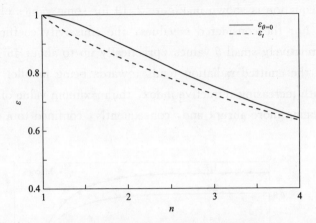

Figure 2.20 Normal and total emissivity coefficients as a function of n (after Astarita and Carlomagno, 2013)

The spectral hemispherical emissivity can be found by integrating Eq. (2.64) over all directions and, as shown in Figure 2.20, the result is particularly interesting. In fact, the hemispherical and normal spectral emissivities are not much different from each other for most practical n values, within less than 10%. This can be very useful when evaluating sensor radiation losses to the ambient environment. Clearly, when using an IR scanner, care should be taken while performing measurements if the scanned object surface is not locally normal to the viewed rays.

From Figure 2.19 it is also evident that, when the viewing angle is relatively small, the directional emissivity of the surface can be correctly assumed to be constant and equal to the normal one. With it being necessary to work at high viewing angles, a careful calibration of $\varepsilon_{\lambda\theta}$ as a function of θ must be performed. Therefore, in order to also take into account the viewing angle variations within the field of view, an optical calibration of the IR scanner should likewise be accomplished in order to correctly estimate the local viewing angles on the test model.

However, while performing experiments with complex model shapes, very high viewing angles should be avoided, because of the steep decrease of $\varepsilon_{\lambda\theta}$ in that range. In such a case and whenever possible, it is much better to take two or more images of the model to be tested from different viewing angles and to reconstruct the thermal image of the model surface (Astarita and Carlomagno, 2013).

2.5.5 Electrically conducting materials

Metals are the most common electrically conducting materials and, in contrast to dielectrics, the extinction coefficient κ is no longer neglectable with respect to the refractive index n, which is much higher. Also in this case, the electromagnetic theory provides a relationship for the evaluation of the spectral directional emissivity but, since it is significantly more complex than Eq. (2.65), the related theoretical will be neglected herein (for details see Siegel and Howell, 1992).

It is interesting to plot, for some $n=\kappa$ values, the spectral directional emissivity coefficient because, as clearly shown in Figure 2.21, its behaviour is different from that of dielectric materials. Usually, the emissivity coefficient of electric conductors has smaller values, with respect to dielectrics, and a a relative minimum in the normal direction. The maximum is obtained for large values of the angle θ and these effects are more pronounced for larger extinction coefficients.

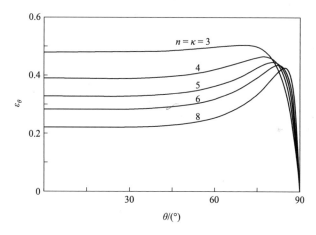

Figure 2.21 Spectral directional emissivity coefficient of conductor materials (after Astarita and Carlomagno, 2013)

The spectral normal directional emissivity can be evaluated with the following straightforward formula,

$$\varepsilon_{\lambda\theta=0} = \varepsilon_{\lambda n} = \frac{4n}{(n+1)^2+\kappa^2} = 1 - \frac{(n-1)^2+\kappa^2}{(n+1)^2+\kappa^2} = 1 - r_{\lambda\theta=0} \quad (2.66)$$

As compared to dielectrics, the significant decrease of the normal emissivity of metals is associated with the higher n values and with the addition of the term κ^2 to the denominator of the first fraction of Eq. (2.65).

Experimental data shows that most metals, unless oxidized or with a rough surface, have a normal emissivity that seldom exceeds 0.2 and very often, if they have a well-polished surface, values can fall below 0.1. Consequently, the perpendicularly emitted radiation is low and the reflected one is high.

This is the reason why, while using IR thermography, metal models are not to be used as such but must be covered with a thin layer of thermally black paint (such as dull enamel) or superficially treated. The former practice cannot be carried out if transient heat transfer is involved.

When transient heat transfer is present, one way to increase surface emissivity is through fine sandblasting of the metal model. The generated superficial roughness acts as a series of microscopic block bodies, so providing a higher emissivity coefficient. Slightly oxidizing the model surface or the use of chemicals may also be appropriate since they exhibit almost the same effect as sandblasting.

Acknowledgements

The following authors, Gaussorgues (1994), Maldague (1993), Astarita and Carlomagno (2013), Halliday et al. (2005), Hudson (1969), and Zhang and Fang (2004), are gratefully acknowledged for the excellent works in their publications directly and indirectly quoted in this chapter.

References

Astarita T, Carlomagno G M. 2013. Infrared Thermography for Thermo-fluid-dynamics. Berlin Heidelberg: Springer-Verlag.
Baehr H D, Stephan K. 2006. Heat and Mass Transfer Berlin: Springer.
Chang S L, Rhee K T. 1984. Black body radiation functions. Int. Comm. Heat Mass Transfer, 11: 451-455.
Dello I G. 2008. An improved data reduction technique for heat transfer measurements in hypersonic flows. PhD Thesis, www.fedoa.unina.it/3039/1/delloioio_2008.pdf, Universita di Napoli.
Gaussorgues G. 1994. Infrared thermography, English language edition translated by S. Chomet. Springer Science+Business Media Dordrecht.
Gebbie H A, Harding W R, Hilsum C, et al. 1951. Atmospheric Transmission in the 1 to 14 μm Region. Proc. Royal Society A, 206: 87-107.
Halliday D, Resnick R, Walker J. 2005. Fundamentals of Physics. 7th ed. Beijing: Higher Education Press.
Hudson R D. 1969. Infrared System Engineering. New York: Springer.

Maldague X P V. 1993. Nondestructive Evaluation of Materials by Infrared Thermography. New York: Springer-Verlag.

Siegel R, Howell J R. 1992. Thermal Radiation Heat Transfer. Washington, DC: Taylor & Francis.

Zhang J Q, Fang X P. 2004. Infrared Physics. Xi'an: XIDIAN University Press.

Chapter 3 Geomechanical model test

3.1 Literature review on physical model test

Due to the dramatic global increase in urbanization, underground caverns have become an attractive option for the construction of powerhouses, transportation conduits, energy storage facilities, and municipal utility systems. A comprehensive understanding of phenomena related to underground excavation, such as deformation patterns and failure mechanisms, is necessary to ensure the safety and continued operation of such facilities (Zhu et al., 2010).

Extensive researches have been conducted on tunneling, roadway excavation and reinforcement, block caving and stability of the underground caverns in sedimentary rocks including, for example, in-situ tests (Li et al., 2008; Tang and Kung, 2009); analytical studies (Lydzba et al., 2003); numerical modeling using finite difference method (FEM) (Tsesarsky, 2012), finite element method (FEM) (Golshani et al., 2007; Fortsakis et al., 2012), discrete element method (DEM) (Heuze and Morris, 2007), discontinuous deformation analysis (DDA) (Hatzor and Benary, 1998; Tsesarsky and Hatzor, 2006; Mazor et al., 2009; Zuo et al., 2009), and geomechanical model tests (Sharma et al., 2001; Kamata and Masimo, 2003; Liu et al., 2003; Castro, 2007; Lee and Schubert, 2008; Shin et al., 2008; Fekete et al., 2010; Zhu et al., 2011; Li et al., 2013).

Among these methods, geomechanical model tests and numerical simulations are most widely used (Zhu et al., 2010). However, as the geological conditions of underground projects become increasingly more complex with the unstable-violent failure modes involved, it is becoming increasingly difficult to represent response using numerical methods with high degree of confidence. Although some discontinuum-based numerical methods (e.g. 3DEC and DDA) can be applied in such cases, they remain innately limited by computational constraints and inadequate understanding of correct physical response of complex media under complex loading conditions. Physical model tests offer a powerful method to explore the influence of material nonlinearities, plastic flow, and spalling failure in strutures constructed in discontinuous rocks. Well designed physical model experiments with judicious choice of model materials may yield important insights into behaviour that are not available from numerical models (Zhu et al., 2011).

As an example, a well-designed large-scale geomechanical model test conducted by Zhu et al. (2008) for stability analysis of a cavern group at great depth is reviewed here. The prototype cavern complex, found in Shuangjiangkou Hydropower station in Sichuan Province, China, contains a power house, a transformer house, a tail water surge chamber. Zhu et al. (2010, 2011) reported the experimental setup using a stiff modular loading frame, hydraulically applied simulated loads and vivo excavation to represent cavern construction at depth. The model experiments are instrumented with fiber Bragg displacement sensors and monitored by digital imaging of fiducial points via an endoscope enabling the effects of rock reinforcement in suppressing failure to be followed.

Figure 3.1 shows schematically the constructed geomechanical model precasting using the analogy material. The physical model has a dimension of 2.5×2 m in horizontal cross section and 2 m in height. Although largely arranged to replicate plane strain conditions, the finite thickness of the model and the inclusion of linking busbar chambers between the main power house and the transformer house violate this precise condition. Consequently the model is considered as a quasi-three-dimensional model.

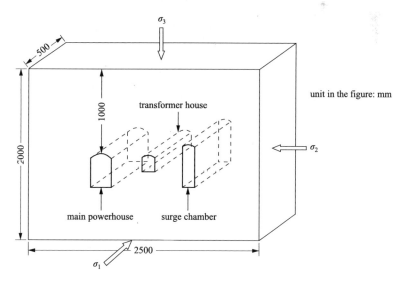

Figure 3.1 Schematic of the geomechanical model (after Zhu et al., 2010)

The structural steel frame is primarily used for such quasi-three-dimensional plane strain model tests with the thickness of the model adjusted within a certain range as shown in Figure 3.2 (Zhu et al., 2010). The frame, accommodating the model, is composed by a base, a door-shaped reaction frame, a layered reaction frame, structural walls, loading jacks, and combination sliding walls, and can be used for stability analysis of underground caverns under high loads equivalent to gravity loads of more than 2000m. High-strength tempered-glass windows are installed on the structural walls around the caverns so that the cracking in the surrounding rock masses can be observed. Uniform and stepwise loading can be applied to the top, base, and left and right sides of the model. The jacks are installed on five sides of the model. For the front and back directions, the layered reaction frame and the corresponding reaction beams can apply self-balanced loading along the axial direction of the model. Many combinational ball-sliding blocks are installed between the model surfaces and the structural walls to reduce friction.

Figure 3.2 Photograph of the steel structural fram for the geomechanical model test
(after Zhu et al., 2010)

Figure 3.3 shows schematically the loading conditions applied to the frame (Zhu et al., 2011). The hydraulic loading system is able to generate non-uniform loads to simulate actual in-situ stress and to simulate a spectrum of loading conditions ranging from low to very high stress. A total of six independent pumps apply the load with small measuring system compressibility. In the frame, total of fifteen hydraulic load cells were installed to generate both gravitational and lateral loads. Five loading system can apply load in each of the three directions, each of which consists of two hydraulic jacks and one hydraulic pressure controlling system. Each hydraulic jack can provide a maximum load of 300 kN, and maintain a consistent load over the loading term (at least 15 days).

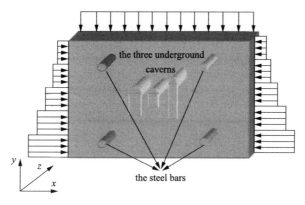

Figure 3.3 Schematic of the loading conditions for the physical model (after Zhu et al., 2011)

Since 2007, systematic experimental studies have been carried out, at the state-key laboratory for geomechanics and deep underground engineering (SKL-GDUE) in China University of Mining and Technology Beijing (CUMTB), on the tunnel excavation and the its stability under high overburden stresses in stratified rock masses based on the large-scale geomechanical model tests and infrared thermography. These physical modeling investigations include the excavation in 0°, 45°, 60° and 90° inclined stratified rocks (He et al., 2010a, 2010b; He, 2011; Gong et al., 2013b) and deformation and failure of the overloaded tunnels embed in horizontally inclined strata (He et al., 2009; Gong et al., 2015). The key and detailed contents will be documented in concerning chapters in this book.

3.2 Similarity theory and dimensional analysis

3.2.1 Similarity principles

The laboratory geomechanical model test must satisfy the similarity principles in order to reproduce the rock behaviour at laboratory with a faithful manner. The similarity principle requires that the model should be analogous to the prototype in terms of geometry, deformation, stresses and strengths, physico-mechanical properties, boundary conditions, and initial stress conditions. The similitude can be derived from the theory of elasticity, i. e,. the equations of equilibrium, compatibility, geometry and physics. The physical entities ratio of prototype (p) to model (m) is called the *similarity constant* or *scale factor* (C). For a model subjected to damaging forces, the stress and strain should be the same as those in the prototype in the elastic range and beyond it until the point of failure; in the post-elastic phase, the residual strains of the model and prototype should be the same (Lin et al., 2015).

The similitude theory about physical modeling experiments requires that some similarity coefficients, defined as the ratios of prototype parameters to model parameters, must be constants (Fumagalli, 1973; Zhu et al., 2011). These prototype parameters are geometry, stress, strain, displacement, deformation modulus, Poisson's ratio, body force, density, friction coefficient, cohesion, compressive strength, and boundary stress. In order to ensure that the physical modeling under the statically increased loads can represent the behaviour of the prototype proportionally with high fidelity, the following conditions must be met, i. e. , similarity of the geometry, similarity of forces and, similarity of the physical and mechanical properties. It is noted that if the dynamical events such as rock burst or coal bump were involved in the physical modeling, the similarity of the time-dependent behaviour is also required.

The three similarity conditions can be defined according to connotation, i. e. , ① for the geometric similarity, the geometric structure of the model should be similar with the prototype; ② for the similarity of the forces, the body stress, external loads, tractions (normal and shear stresses) on the surface, body force, boundary stresses, and compressive and tensile strength of the model should be similar with the prototype, as well as the stress-strain relationship curve; and ③ for similarity of the physical properties, the deformation modulus, density, friction coefficient, cohesion, and Poisson's ratio of the model should be similar with the prototype.

The geometry *similarity constants* (or *scale factors*) are given by,

$$C_l = l_p/l_m, \ C_\varepsilon = \varepsilon_p/\varepsilon_m, \ C_\delta = \delta_p/\delta_m \qquad (3.1)$$

where, C_l, C_ε and C_δ represent the similarity constants for characteristic length, strain and displacement, respectively. The subscript p denotes that the corresponding parameter is of prototype, while the subscript m indicates that the corresponding parameter is that of the physical models.

The force similarity constants include,

$$C_\sigma = \sigma_p/\sigma_m, \ C_R = R_p/R_m, \ C_T = T_p/T_m, \ C_X = X_p/X_m, \ C_{\bar\sigma} = \bar\sigma_p/\bar\sigma_m \quad (3.2)$$

where, C_σ is the stress scale factor; C_R is the compressive strength factor; C_T is the tensile strength factor; C_X is the body force factor; and $C_{\bar\sigma}$ is the boundary stress factor.

The physical property similarity constants are,

$$C_E = E_p/E_m, \ C_\nu = \nu_p/\nu_m, \ C_\rho = \rho_p/\rho_m, \ C_f = f_p/f_m, \ C_c = c_p/c_m \quad (3.3)$$

where, C_E is the similarity constant of the deformation modulus; C_ν is the similarity constant of the Poisson's ratio; C_ρ is the density scale factor; and C_c is the cohesion scale factor.

In the case of shear sliding being the major mode of the failure, the similarity constant of shear strength, $C_\tau = \tau_p/\tau_m$, also needs to be defined. But in a general situation, the shearing similarity can be derived from the similarities of the cohesion and friction as defined in Eq. (3.3). Moreover, if the heat transfer is involved, the thermodynamics similarity should also be considered.

According to the physical and geometrical relations, the following equations are valid,

$$C_\gamma = (\rho g)_p/(\rho g)_m = C_\rho \quad (3.4)$$

where, C_γ is the *gravitational similarity (body force) constant* which is equal to that of the density, C_ρ; g is the acceleration of gravity; and,

$$C_\delta = C_\varepsilon C_l \quad (3.5)$$
$$C_\sigma = C_E C_\varepsilon \quad (3.6)$$

The following similarity criteria must also be satisfied in physical modeling (Fumagalli, 1973; Zhu et al., 2011):

$$\frac{C_\sigma}{C_\rho C_l} = 1 \quad (3.7)$$

$$\frac{C_\delta}{C_\varepsilon C_l} = 1 \quad (3.8)$$

$$\frac{C_\sigma}{C_\varepsilon C_E} = 1 \quad (3.9)$$

$$C_\nu = C_\varepsilon = C_f = 1 \tag{3.10}$$

If the similarity constant for density is determined as $C_\rho=1$, hence $C_\gamma=1$, the other similarity constant can be obtained according to Eq. (3.7) to Eq. (3.10) (Zhu et al., 2011). These four equations are the guiding parameters for choosing or manufacturing the modeling materials of the physical model tests with specified mechanical properties, model size and proportions, magnitudes of loading forces, and boundary conditions (Fumagalli, 1973; Zhu et al., 2011).

3.2.2 Selection of similarity materials and ratios

The Eq. (3.4) and Eq. (3.9) constitute the guiding principles for selecting or fabricating the modeling materials of the physical model tests with specified mechanical properties, model size and proportions, magnitudes of loading forces, and boundary conditions. According to Eq. (3.9) and Eq. (3.10), the target experimental materials must be high density, low Young's modulus, low strengths (Da Silveira et al., 1979).

Liu et al. (2013) reviewed some sample materials widely used in the geomechanical model test.

(1) PbO or Pb_3O_4 as aggregate, ZnO as supplement (Fumagalli, 1973; Lemos, 1996). The material has high density. However, PbO or Pb_3O_4 is poisonous and ZnO is expensive. Some research used lithopone (Jiang et al., 2009) as a substitute for ZnO, the result is satisfactory.

(2) Mixture of epoxy resin, barite powder and glycerin (Shen and Zou, 1988). This material has appropriate strength and Young's modulus. The disadvantate is that the material needs to be solidified by high temperature, during which poisonous gas is generated as well.

(3) Fluid wax as binder. These materials are mechanism property stable, not sensitive to temperature or humidity changing. Besides, the pressed blocks need not to be dried. However, the fluid is expensive. Some researchers used engine oil as substitute, the result is satisfactory (Shen and Zou, 1988).

(4) Gypsum materials, sand or diatomite as supplement (Gong et al., 1984; Oliverira and Faria, 2006). It is not easy to handle with the water proportion and solidifying time for these materials.

(5) High density metal materials, such as iron (Han et al., 1983), lead (Du, 1996), and copper (Gu et al., 1984) powder as aggregates. The disadvantage is obvious: they are expensive. In addition, lead is poisonous, iron is

constringent and rusty, and appropriate copper powder is not easily found.

(6) NIOS (Ma et al., 2004) or IBSCM (Wang et al., 2006). These materials perfectly meet the similarity requirement. The disadvantage is that the drying process of NIOS is slow, while IBSCM is expensive by containing the refined iron ore.

(7) Temperature-analogue material (Fei et al., 2010). These materials contain fusible macromolecular, which could simulate the strength reducing process.

The recent development on the modeling materials include:

(8) Barite powder as aggregate, bentonite as supplement, and glue as binder (Liu et al., 2013). Barite has high density. Bentonite is stable and inexpensive. Moreover, it could be decreased the mixtures Young's modulus. Glue is useful for simulating the low strength of material. This analog material has several advantages. It satisfies all of the similitude's requirements, is inexpensive, and suitable for simulation of brittle materials, such as concretes and rocks. As a result, it well balanced the technical and economic considerations.

(9) New type of modeling material being developed, comprising of iron, barite, and quartz powders bound with a solution of alcohol and rosin-the alcohol evaporates and leaves the rosin as a binder (Zhu et al., 2011).

(10) Similar materials used to simulate rock masses with different friction coefficients f and cohesions c, which are made of barite powder, sand, expansion soil, and water with various proportions. For the type Ⅲ rock mass, the proportion of the barite powder, sand, expansion soil, and water in the modeling material is 9.5 : 3.5 : 0.3 : 1 (Lin et al., 2015).

In the most cases in selection of the analog materials listed above, the similitude relation, $C_\gamma=1$ is usually required, so as to keep the test normal (Liu et al., 2013). Then according to Eq. (3.4), $C_\gamma=C_\rho=1$, thus simulation of the self-weight can be done by letting the density of the modeling materials equal to that of the corresponding prototype materials. Then Eq. (3.7) can be rewritten as,

$$C_\sigma = C_\rho C_l = C_\gamma C_l \tag{3.11}$$

According to the similarity principles, the force scale should be equal,

$$C_E = \frac{E_p}{E_m} = C_\sigma = \frac{\sigma_p}{\sigma_m} \tag{3.12}$$

where,

$$E_m = \frac{E_p}{C_\sigma} = \frac{E_p}{C_\gamma C_l}$$

Thus the following relation will be valid,

$$C_E = C_\rho C_l = C_\gamma C_l \tag{3.13}$$

Under the condition $C_\gamma = 1$, the gravitational load could be well applied, and the conversion of other parameters could be straightforward. For example, from Eq. (3.13) we can obtain,

$$C_E = C_l \tag{3.14}$$

3.3 Field case (prototype)

3.3.1 Site geology

The field case (prototype) simulated in this research is a main haulage roadway under operation in QISHAN underground coal mine, located in Xuzhou mining district, Jiangsu province, eastern China. The mining depth of this mine at present is ranging from 300-1000 m and will proceed to greater depth of more than 1000 m in the future. Note that the term *"depth"* in the following context is referred to the vertical length counting from the ground surface. The coal seam is inclined at different inclination angles ranging from 0° to 90° with respect to the horizontal, with black colored, semi-bright coal rocks in the form of block and granular. Geological formation of the mine is composed by the coal seam between successive layers of sandstone and mudstone.

Rocks adjacent to the roadway are composed mainly by mudstone with coal and sandstone distributed in some localized sites. Field observation of the outcrop showed that joints and fractures in the surrounding rock masses are well developed; under the influences of the faults, the strata are generally fractured. The haulage roadway tunnel strikes NE6°. The ground stress measurement showed that the maximum principal stress at -1000 m level is 40.5 MPa, and the azimuth angle is 140°. Therefore, the angle between the striking of the roadway and the maximum principal stress is 46°, unfavorable for roadway support.

The geological structure of QISHAN coal mine is complexed, distribution of the fracturing and folding zones manifests the result of the superposition effect of the stresses in different directions with different magnitude levels. The stratigraphic column of the surrounding rocks is given in Figure 3.4. It is seen from the figure that the surrounding rocks can be classified into three engineering rock

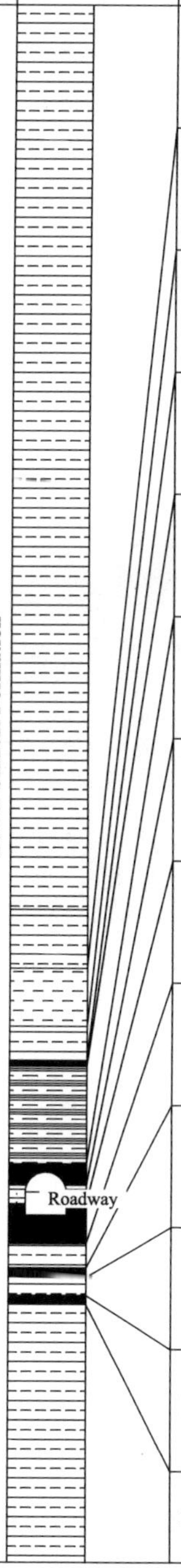

Stratigraphy unit	Column	No.	Thickness /m	Accumulated/m	Rock type	Color	Rock property discription
The upper Paleozoic / The Permian / Two Triassic / Lower Shihezi Formation		1	77.18	77.18	mudstone	motley	Pelitic texture, compactation, flat fracture, containing fern fossils, variegated and gray-green colored, and with sliding surfaces
		2	3.8	80.98	sandstone with moderate sized grains	gray white	Pelitic texture, compactation, flat fracture, containing fern fossils, variegated and gray-green colored, and with sliding surfaces
		3	2.1	83.08	mudstone	gray black	Pelitic texture, compactation, flat fracture, containing fossil plant fragments
		4	0.2	83.28	coal seam	black	Sublayer of the #1 coal seam
		5	6.3	89.58	sandy mudstone	gray black	Pelitic texture, compactation, flat fracture, brittle, and containing fossil plant fragments
		6	1.4	90.98	#1 coal seam	black	Lumped, granular, and semi bright
		7	1.2	92.18	mudstone	deep gray	Logging data
		8	2.5	94.68	#3 coal seam	black	Lumped, granular, and semi bright
		9	1.53	96.21	mudstone	deep gray	Logging data
		10	0.65	96.86	#4 coal seam	black	Logging data
		11	1.36	98.22	mudstone	deep gray	Logging data
		12	0.56	98.78	coal seam	black	Logging data
		13	16.88	115.66	mudstone	motley	Motley, politic texture, flat fracture surface, containing fossil plant fragments and ling iron grains

Figure 3.4 Generalized stratigraphic column at QISHAN coal mine, Xuzhou mining district

groups, i.e., the sandstone group (including moderate grained stand stone and fine grained stand stone), mudstone group (including mudstone and sandy mudstone), and coal seam group (including #1, 2 and 3 coal seams). The roadway tunnel was excavated in the #3 coal seam. Severe damages to the roadway were encountered at QISHAN coal mine both in the early development phase and operational phase, including large deformation of the surrounding rocks and support system. The roadway failure includes the following aspects: ①deformation of the two-side walls are asymmetric and non-uniform (see Figure 3.5a); ②significant floor heave was encounter during the roadway development (see Figure 3.5a); and ③(see Figure 3.5b) different degrees of roof subsidence, collapse and sprayed layer peeling off were encountered during the operational phase.

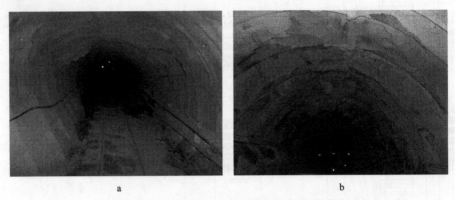

Figure 3.5 Roadway failure at QISHAN coal mine
a. asymmetric deformation of the two-side walls and floor heave;
b. roof collapse with around 300 m³ rock collapsed

3.3.2 In situ rock properties

According to the field survey, three rocks, i.e., sandstone, mudstone and coal are the major rock types found in the vicinity of the roadway at QISHAN. As a matter of fact, the surrounding rock masses at depth are mainly composed by the three rock types and their combination in space. The geomechanical models, therefore, were designed to be constructed with alternating layers of sandstone, mudstone and coal seam to simulate the surrounding rock masses depicted in Figure 3.4.

Rock samples of sandstone, mudstone and coal were cored by drilling in the haulage roadway at -1000 m level at QISHAN coal mine. Basic physical and mechanics tests, including the uniaxial compressive tests, Brazilian disc tests, triaxial tests, were conducted at laboratory. The physical and mechanical parameters of the real rocks are reported in Table 3.1, which are the real rock properties simulated using the artificially produced materials. It is seen that the UCS (unconfined compressive strength) for the surrounding rocks are generally low, especially for the mudstone (43.78 MPa).

Table 3.1 Physical and mechanical parameters for the real rocks

	Unit weight /(kN/m³)	UCS /MPa	Tensile strength /MPa	Young's modulus /GPa	Internal friction angle /(°)	Poisson's ratio	Cohesion /MPa
Sandstone	26.55	63.98	5.83	25.77	33.71	0.15	16.51
Mudstone	25.78	43.78	5.59	21.01	36.35	0.13	23.59
Coal seam	14.0	26.15	0.90	5.0	40.07	0.36	5.42
Analogue ratio (sand : mud : coal)	1.9 : 1.9 : 1	2.5 : 1.7 : 1	6.5 : 6.2 : 1	5.2 : 4.2 : 1	0.9 : 0.9 : 1	0.4 : 0.4 : 1	

3.4 Geomechanical model construction

3.4.1 Testing machine

The test machine used in the large-scale geomechanical model tests introduced in this book is the so-called "YDM-C Geological Disaster Simulation Testing Machine", as shown in Figure 3.6. The steel frame of the testing machine can accommodate a model of 1600 mm×1600 mm×400 mm (length×height×width) and apply scaled loads as much as 7 MPa with an accuracy less than 1‰. The testing machine can be used for performing the quasi-two-dimensional large-scale geomechanical model test. It consists of a gate-shaped steel frame, loading platens, hydraulic loading-control device, and data collection system.

The frame is composed by four beams including the right and left stands and the top and lower beams which are connected by the screw rods, cap nuts and junction plates imposing uniform stresses on the boundaries of the geological model. The top beam and the right and left stands were installed with six loading platens respectively for imposing uniform stresses on the boundaries of the physical model. The loading platens on the top beam are used to generate gravi-

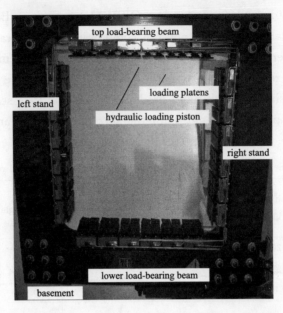

Figure 3.6   Photograph of the Testing frame

tational load, and the loading platens on the two side stands are used to apply lateral loads. Each of the hydraulic-driven-controlled loading platens can maintain a consistent load over the long term. The stress/strain field is measured by the data collection system which consists of the strain gauges, data collecting module.

3.4.2   Model dimension

The similarity principle in the geomechanical model test means that the physical features of the model should be analogous to those in the phototype cases including the modeling material, loads, shape and dimension. For a linear elastic model, the similarity can be obtained from the theory of elasticity, as introduced and formulated previous sections.

For a model subjected to damaging forces, the stress and strain should be the same as those in the prototype in the elastic range and beyond it until the point of failure; in the post-elastic phase, the residual strains should be the same as $\varepsilon_p^0 = \varepsilon_m^0$, where ε^0 is the residual strain; and each point inside the model must follow the equilibrium, compatibility, and geometrical equations. The points on the model surface should satisfy the boundary conditions (Lin et al., 2014; Lin et al., 2015).

In addition to the body force factor, another controlling factor to choose is the geomechanical factor C_l in the physica model test. As indicated by its definition, $C_l = l_p/l_m$, the geomechanical factor links the scale of the engineering rock mass to the scale of the physical model limited by the laboratory testing frame. In the presented cases of physical models tests in this book, the simulated tunnel excavation has a cross-sectional dimension of 3 m×2.4 m, and the excavation influenced zone occupies an area around 20 m×20 m in plane dimension. The maximum plane dimension of the frame containing the physical model is 1.6 m×1.6 m.

When the geometrical factor $C_l \geqslant 12$, the tunnel cross-section will be too small to allow a faithful simulation of the in situ excavation. At the same time, the monitoring devices will be difficult to install and the monitored physical parameters will be inaccurate. On the contrary, if the geometrical factor is smaller than 12, then the sizes of the cavern will be larger than the limiting dimension of the testing frame. Based on these considerations, the optimal value for the geometrical factor is estimated as,

$$C_l = \frac{l_p}{l_m} = 12 \qquad (3.15)$$

The dimension of the geomechanical model could be determined using the geometrical scale factor as shown in Figure 3.7. The overall dimension of the model that the frame contains is 1.6 m×1.6 m×0.4 m, corresponding to an engineering rock mass (in situ) with cross-section area 19.2 m×19.2 m. Dimension of the cross-section for the excavated tunnel is 0.25 m×0.2 m (length × height), corresponding to an underground tunnel with a cross-sectional area of 3 m×2.4 m.

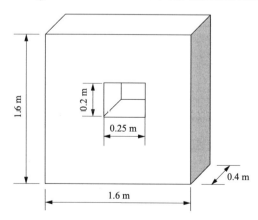

Figure 3.7   Geometry of the geomechanical model

3.4.3 Physico-mechanical parameters of the model

In addition to the geometrical similarity constant C_l, another important similarity constant is stress scaling factor C_σ which can be estimated based on the maximum boundary stress and loading capacity of the frame. The maximum loading capacity of the frame is 7 MPa (σ_m, sub-section 3.4.1). Considering the planning mining will be performed at a depth of 2000 m or more. The unit gravitational force of the surrounding rocks, according to the stratigraphy in Figure 3.4, can be generalized as 27 kN/m³. Thus the maximum vertical stress equals to,

$$\sigma_p = \gamma H = 27 \times 2000 = 54000 \text{MPa or } 54 \text{MPa}$$

Then, the stress scaling factor is,

$$C_\sigma = \sigma_p / \sigma_m = 54/7 = 7.71 \approx 8$$

By using similarity constants C_l and C_σ, the body force scale factor can be determined,

$$C_\gamma = \frac{C_\sigma}{C_l} = \frac{8}{12} = 0.67$$

The ratio between elasticity modulus is,

$$C_E = \frac{E_p}{E_m} = C_\sigma = 8$$

The strain ratio,

$$C_\varepsilon = \frac{\varepsilon_p}{\varepsilon_m} = 1$$

Poisson's ratio,

$$C_\nu = \frac{C_p}{C_m} = 1$$

The internal friction angle scaling factor,

$$C_\phi = \frac{\phi_p}{\phi_m} = 1$$

Gypsum has been recognized as a good rock model material and a significant laboratory experiments on crack initiation and coalescence were conducted with this material (Bobet and Einstein, 1998; Sagong and Bobet, 2002; Mutlu and

Bobet, 2006). Using gypsum with different ratios of water to gypsum, different rock types can be produced having similarities in terms of the major mechanical and physical properties. The most prominent advantage for using the water-gypsum modeling material lies in the easy-to-get, low costs and suitable for producing a large quantity of the modeling materials in the large-scale geomechanical model tests.

Three modeling rock materials used in the geomechanical model tests involved in this book were produced with three water to gypsum ratios, among them, 0.8 : 1 for sandstone, 1:1 for mudstone and 1.2 : 1 for coal. By using the similarity constants determined above and the physico-mechanical parameters of real rocks (Table 3.1), the physio-mechanical parameters of the three modeling materials can be readily obtained (see Table 3.2). For evaluation of the analogous degree of the analogue and real materials, we defined a factor, analogous ratio (AR, AR=sand : mud : coal), to compute the relative ratios for each of the material and mechanical parameters for the real rocks and model rocks (see Table 3.1 and Table 3.2).

Table 3.2 Mechanical, material and geometric parameters of the artificial materials used in the physical model

Rock types simulated by the elementary slab	Dimensions of the elementary slabs/cm	Unit weight /(kN/m³)	Unconfined compressive strength /MPa	Tensile strength /MPa	Young's modulus /GPa	Internal friction angle/(°)	Poisson's ratio
Sandstone	40×40×3	15	8	0.72	3.22	32	0.13
Mudstone	40×40×2	11	5	0.69	2.62	33	0.12
Coal seam	40×40×1	8	3	0.11	0.61	33	0.32
Sandstone	40×40×3	15	8	0.72	3.22	32	0.13
Mudstone	40×40×2	11	5	0.69	2.62	33	0.25
Analogous ratio (sand : mud : coal)		1.9 : 1.4 : 1	2.5 : 1.7 : 1	6.5 : 6.3 : 1	5.3 : 4.3 : 1	0.9 : 1 : 1	0.4 : 0.4 : 1

3.4.4 Rock structure simulation

The surrounding rocks of the prototype are characterized by complicated engineering geology, developed joints and fissures, and very poor stability. For simulating such structures, a compound reference rock mass model can be used based on the GSI rock mass characterizations system and rock mass structure and discontinuities surface quality (Marinos and Hoek, 2000; Marinos et al., 2005;

Fortsakis et al., 2012), as shown in Figure 3.8. The reference rock mass is composed by the dominant discontinuities (e.g. beddings) and internal rock mass without the persistent discontinuities but contains all the secondary discontinuities.

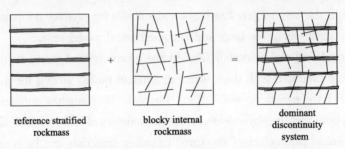

reference stratified rockmass   blocky internal rockmass   dominant discontinuity system

Figure 3.8   Schematic of rock structure stimulated by the geological model
(after Fortsakis et al., 2012)

The reference rock mass can be considered as the qualitative sum of the internal rock mass and dominant discontinuity system. The dominant discontinuities are bedding, schistosity and weak surface; the blocky internal rock mass are often interlocked, partially disturbed and undisturbed rock mass consisting of cubical blocks formed by the orthogonal intersecting discontinuity sets, namely the secondary discontinuities (Fortsakis et al., 2012).

Generally speaking, construction of the large-scale physical model is very expensive and difficult to operate. To overcome these problems, a new approach, i.e., the physically finite elementary slab assemblage (PFESA) was used for constructing the geomechanical models (He et al., 2009). The elementary slab is a specimen-sized prismatic plate fabricated with an intimate mixture of gypsum and water casting against a mold. All the elementary slabs were made with the same dimension of 400 mm×400 mm and thickness of 10, 20 and 30 mm respectively, as reported in Table 3.2. Figure 3.9 shows the produced elemental slabs in our laboratory for construction of the rock strata.

Figure 3.10 shows the principle for producing rock strata. A large number of the elementary slabs (shown in Figure 3.9) with the same model-rock property were used to ensemble a rock layer of the same property and a certain number of the rock layers with the same rock property were used to construct a rock stratum. As shown schematically in Figure 3.10, the mudstone stratum and coal seam stratum are assembled by placing the elementary slabs in layers with perfected mated interfaces.

Chapter 3　Geomechanical model test · 67 ·

Figure 3.9　Photograph showing the manufactured elemental slabs

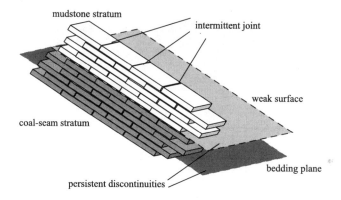

Figure 3.10　Model rock strata construction

A weak surface is formed between the two different model rock strata (see Figure 3.10). This weak surface is homogeneous along its striking at macroscopic scale while has many minor asperities at "microscopic scale" constituted by many small gaps between the parallel-placed elemental slabs in the rock layer. The gaps and asperities formed by the layered slabs also exist within the rock strata. These minor flaws can be viewed as the origin of the heterogeneity of the rock under external loading. Three classes of the model-rock strata were constructed in our test, i. e., sandstone stratum, mudstone stratum and coal stratum which create two types of interfaces or weak faces, i. e., sandstone-coal interface and mudstone-coal interface.

3.4.5 Geomechanical model

The geomechanical models can be built with alternating layers of the modeling rock strata of sandstone, mudstone and coal seam. For instance, Figure 3.11 shows schematically the constructed geomechanical model of 45° inclined rock strata. The model is 1.6 m high and 1.6 m wide and 0.4 m in thickness. The excavation zone was located in the center of the model within the stratum 4 (coal seam) and has a dimension of 200 mm in height, 250 mm in width and 400 mm long (equal to the model's thickness). During the roadway excavation or overloading, different boundary conditions can be given using the testing frame (will be described in the following chapters).

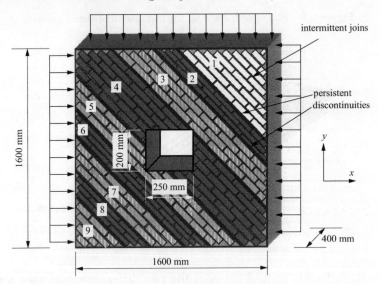

Figure 3.11 Schematic of the constructed geomechanical model of 45° inclined rock strata
(Gong et al., 2015)

It is seen from the Figure 3.11 that the model is composed by total nine strata including one sandstone, four mudstones and four coal seams. These rock strata are indexed with digits from 1 to 9 from the left to the right. The geological section of the physical model is shown in Figure 3.12, including the thickness of the strata and the number of layers of the elementary slabs used in each stratum assemblage.

Stratum No.	Thickness /mm	Rock type	
1	440	sandstone	
2	140	coal seam	
3	120	mudstone	
4	250	coal seam	
5	150	mudstone	
6	60	coal seam	
7	140	mudstone	
8	60	coal seam	
9	240	mudstone	

Figure 3.12 Geological section of the constructed geomechanical model

3.5 Infrared detection

3.5.1 Thermography and imaging procedures

Infrared imaging can be carried out during the laboratory tests. For example, when conducting the unsupported roadway excavation experiments using blasting and drill methods, the tunneling can be performed at the back side of the model while the infrared camera is placed in the front side of the model, for capturing the thermal responses of the rock under the excavation impact. As an example, the roadway excavation experiment incorporating with the thermal vision is shown schematically in Figure 3.13. The thermal sequence are captured by thermography and displayed in real time. At the same time, the acquired thermal sequences are stored in the imager as digital images which are processed using the image processing technique for detecting the rock behaviours.

As discussed in Chapter 2, for detecting the thermal behaviour from the geomechanical tests at laboratory condition, suitable working band for the thermography is the *middle infrared* (3-6 μm) band. Many commercial models of the thermography can be found in the market, such as the thermal imagers produced by manufacturers of Infrared Tec, FLUKE and FLIR, etc.

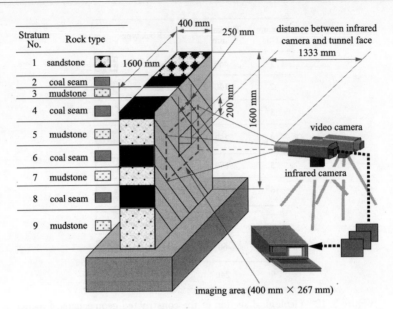

Figure 3.13 Thermal imaging of the geomechanical model test (Gong et al., 2015)

In the following, the usage and technical performances of the thermography TVS-8100 MK Ⅱ, employed in the practical cases of the infrared detection introduced in this book, will be introduced. The Thermal Video System, TVS-8100 MK Ⅱ Series, is a thermal imaging system with radiometric temperature capability utilizing 2-dimensional infrared detector. The fine image is acquired at 60 frames per second (Nippon Avionics Co. Ltd., 1990).

The TVS consists of two main component parts: the Imager (or infrared camera head) and the processor as shown in Figure 3.14. The imager will receive infrared energy emitted from the surface of object and convert this into an electrical signal. The imager includes 2-dimensional infrared detector without mechanical scanning, therefore speed thermal image or small changing thermal image is acquired to display on the monitor screen at real time. The processor will store the electrical signal from the Imager in frame memory and process it. When a complete frame of infrared information has been acquired, it is displayed at 256 colors on the built-in color LCD monitor in real time.

The TVS offers several built-in image processing and analysis features such as: image averaging, selectable number of displayed levels, display freeze, multi-point temperature indication, as well as time, date and user message. The TVS can also automatically track temperature and record or replay thermal image using

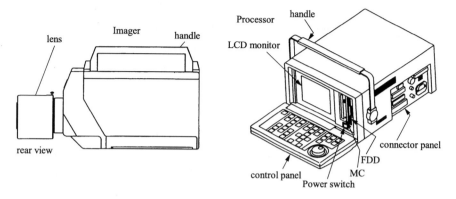

Figure 3.14 Main component parts of the TVS-8100 MKⅡ system, i. e., imager and processor

the built-in floppy disk. However, these "in-situ" image processing is preliminary. With use of peripheral equipment, thermal image may be recorded on file for further processing procedures (will be introduced in the following).

The TVS system include the following components:
(1) Imager (attached 25 mm remote focus lens);
(2) Processor;
(3) Cable, Imager to processor;
(4) Cable, AC 3 wire to 2 wire;
(5) 3.5 inch floppy disk (2HD type, 1.44MB);
(6) Tripod.

The most important useage precaution is to avoid using or storage the system at the following places, i. e., place exposed to direct sunlight or near a heater; humid or dusty place; place where vigorous vibration occurs, and places near such an intense magnetic source as an electric motor.

The infrared detector in the Imager of TVS-8100MKⅡ is cooled by a built-in refrigerator, known as Stirling cooler. A stirling cooler consists basically of a motor assembly, a compressor, and a cold finger. The cooler is filled with helium gas of 99.999% purity or higher. In compressor is a piston, and this piston is driven in a reciprocating manner by the motor. The action of the piston subjects the helium gas to adiabatic expansion and compression, thus cooling the cold finger which is on the compressor's front surface. The infrared detector is attached on the front end of the cold ginger and kept cool by it.

Since mechanical action takes place in the Stirling cooler, parts are subject to wear and tear and the helium gas eventually is contaminated. The contaminated helium gas affects cooling capacity of the Stirling cooler and the TVS performance. Therefore, it is necessary to maintain the Stirling cooler properly to keep the TVS performance. The maintenance interval is 2000 hours. The internal helium must be replaced at every maintenance interval.

The infrared camera works at wave length of 3.6-4.6 μm, with measuring temperature range of -40 to $+300$°C; minimum detection temperature difference of 0.025°C; a field of view of 13.6°×18.2°/25 mm; spatial resolution of 2 mrad; on-line display resolution 240×320 pixels. The raw thermogram were stored in the recorder as digital image of 120×160 pixels for off-line processing.

3.5.2 Temperature calibration

The methodology on the use of infrared thermography, in general, has two categories: i.e., passive thermography and active thermography. Active thermography uses an external heating device for heating up the testing object under external loading and, therefore, needs to calibrate the obtained temperature increase physically by manned-controlled operation with respect to the 'known heat source'.

Passive thermography detects temperature rise of the testing object under external loading without the use of any extra heat sources. When using active thermography, quantitative analysis of thermal image could be performed by the physical calibration of the temperature increment against the applied loads or displacement. In contrast, analysis would be qualitative when using passive thermography. The choice of the modes for using the infrared thermography is related to the geometrical and physical properties of the object in view.

By heating up the object under testing, active thermography can acquire thermal image with large temperature increment. Thus the thermal image has a large dynamic scale and high contrast capable of representing the physical process clearly. Metals and composites generally have a relatively simple constitutive relationship and high thermal conductivity. In this case, the use of infrared thermography in the active mode is preferred.

Active thermography has been widely used but not limited to the cases, for example, damage characterization of the stressed carbon fiber reinforce polymers (Steinberger et al., 2006; Mayr et al., 2011); defect detection of magnetic specimens (Lahiri et al., 2015) and aluminum specimens (Pastor et al., 2008), and quantitative analysis of plastered mosaics (Theodorakeas et al., 2014). However, the active thermography was rarely used for detection of rock materials due to the structural heterogeneity and small thermal conductivity. Thus, when detecting rock materials, the infrared thermography is usually used in a passive mode.

When using passive thermography, although no requirement for the physical calibration of the temperature increment, however, finding a reference point mathematically is needed for characterization of the detected temperature variation, which could be referred to 'mathematical calibration'. The mathematical calibration can be realized by the image subtraction algorithm. While the object under detection is subjected to the external loading, the interested features of the infrared sequence are the temperature increment relative to that of the initial state. Taking the first frame of infrared sequence when the object was at the initial state, subtraction of the first frame from the following images obtains the temperature increment relative to the initial state of the object.

The mathematical calibration was implemented by the image subtraction

$$\hat{f}_k(x,y) = f_k(x,y) - f_0(x,y) \quad (3.16)$$

where, $f_k(x,y)$ represent the image matrix of kth frame in the sequence which is actually the detected temperature field at the kth instant of time; $\hat{f}_k(x,y)$ is the incremental temperature field (also image matrix) at the kth instant obtained by subtraction of the first frame from $f_k(x,y)$; $f_0(x,y)$ is the first frame of the thermal sequence taken at the initial state of the loading; the subscript k is an integer served as frame index; $x=1,2,\cdots,N$ and $y=1,2,\cdots,M$ are the pixel coordinates, and $M=160$ and $N=120$ are the maximum pixel number respectively for the image matrix.

3.5.3 Image processing

The tasks for processing thermal image acquired in the large-scale geomechanical model tests include ① removal of different types of noises, and ② enhancement of the low-contrast image. The image processing techniques were well established in our previous works (Gong et al., 2013b) and the related algorithms used in this research are summarized in the following.

(1) For removal of the environmental radiation noise, the image subtraction expressed in Eq. (3.16) was employed. The image subtraction can also be used as the temperature calibration procedure as introduced in section 3.2.

(2) For eliminating the salt-and-pepper noise induced by the electronic current in the measurement instruments, median filter was used.

(3) For reduction the additive-periodical noise which may come from the rotating parts in the cooling system embedded in the infrared camera, Gaussian high-pass filter (GHPF) in the frequency domain was utilized.

(4) When detecting a large-scale object with the IR camera working in the passive mode, the raw thermal image will have a small dynamic scope. As a result, the images should be enhanced in order to represent the rock response clearly. In this paper, a morphological enhancement filter, κ_n, proved to be very effective and developed by Gong et al. (2013a), was used.

The imaging processing is an object orientated problem and the concerning algorithms may vary from case to case. Detailed discussion on the image process algorithms for the treatment of the low-contrast and noisy thermal images will be given in the following chapters when introducing the cases of the applications of the thermography in geomechanics tests.

References

Bobet A, Einstein HH. 1998. Fractare coalescence in rock-type materials under uniaxialand biaxial compression. International Journal of Rock Mechanies and Mining Sciences. 35(7): 863-888.

Castro R, Trueman R, Halim A. 2007. A study of isolated draw zones in block caving mines by means of a large 3D physical model. International Journal of Rock Mechanics and Mining Sciences, 44(6): 860-870.

Da Silveira A F, Azevedo M C, Esteves Ferreira M J, et al. 1979. High density and low strength materials for geomechanical models. In: Proceedings of the international colloquium on geomechanical model. Bergamo, Italy: ISRM: 115-1310.

Du Y J. 1996. Research present situation and development treng of geomechanics model test. NW Water Res. Water Eng. , 7(2): 64-67(in Chinese).

Fei W P, Zhang L, Zhang R. 2010. Experimental study on a geo-mechanical model of a high arch dam. International Journal of Rock Mechanics and Mining Sciences, 47: 299-306.

Fekete S, Diederichs M, Lato M. 2010. Geotechnical and operational applications for 3-dimensional laser scanning in drill and blast tunnels. Tunnelling and Underground Space Technology, 25: 614-628.

Fortsakis P, Nikas K, Marinos V, et al. 2012. Anisotropic behavior of stratified rock masses in tunneling. Eng. Geol. , 141-142: 74-83.

Fumagalli E. 1973. Statical and Geomechanical Models. New York: Springer, Wien.

Golshani A, Oda M, Okui Y, et al. 2007. Numerical simulation of the excavation damaged zone around an opening in brittle rock. International Journal of Rock Mechanics and Mining

Sciences, 44: 835-845.

Gong S X, Guo C M, Gao D S. 1984. The experimental studies of the geomechanical model material. J. Yangtze River Sci. Res. Inst., 1: 32-46(in Chinese).

Gong W L, Gong Y X, Long A F. 2013a. Multi-filter analysis of infrared images from the excavation experiment in horizontally stratified rock. Infrared Physics and Technology, 56: 57-68.

Gong W L, Peng Y Y, Wang H, et al. 2015. Enhancement of low-contrast thermograms for detecting the stressed tunnel in horizontally stratified rocks. International Journal of Rock Mechanics and Mining Sciences, 74: 69-80.

Gong W L, Wang J, Gong Y X, et al. 2013b. Thermography analysis of a roadway excavation experiment in 60° inclined stratified rocks. International Journal of Rock Mechanics and Mining Sciences, 60: 134-147.

Gu Z Q, Peng S Z, Li Z K. 1984. Underground Chamber Engineering. Beijing: Tsinghua University Press.

Han B L, Zhang W C, Yang C F. 1983. A new model material of geomechanics (MIB). J. Wuhan Univ. Hydraul. Electron. Eng., 1: 11-17(in Chinese).

Hatzor Y H, Benary R. 1998. The stability of a laminated Voussoir beam: Back analysis of a historic roof collapse using DDA. International Journal of Rock Mechanics and Mining Sciences, 2(12): 165-181.

He M C. 2011. Physical modeling of an underground roadway excavation in geologically 45° inclined rock using infrared thermography. Eng. Geol, 121(3-4): 165-176.

He M C, Gong W L, Li D J, et al. 2009. Physical modeling of failure process of the excavation in horizontal strata based on IR thermography. Int. J. Min. Sci. Technol, 19: 0689-0698.

He M C, Gong W L, Zhai H M, et al. 2010a. Physical modeling of deep ground excavation in geologically horizontally strata based on infrared thermography. Tunnelling and Underground Space Technology, 25: 366-376.

He M C, Jia X N, Gong W L, et al. 2010b. Technical note, Physical modeling of an underground roadway excavation vertically stratified rock using infrared thermography. International Journal of Rock Mechanics and Mining Sciences, 47: 1212-1221.

Heuze F E, Morris J P. 2007. Insights into ground shock in jointed rocks and the response of structures there-in. International Journal of Rock Mechanics and Mining Sciences, 44: 647-676.

Jiang X L, Chen J, Sun S W, et al. 2009. New material for rock foundation of geomechanical model of Jingping Project. J. Yangtze River Sci. Res. Inst., 26(6): 40-43(in Chinese).

Kamata H, Mashimo H. 2003. Centrifuge model test of tunnel face reinforcement by bolting. Tunnelling and Underground Space Technology, 18(2-3): 205-212.

Lahiri B B, Bagavathiappan S, Sebastian L T, et al. 2015. Effect of non-magnetic in magnetic specimens on defect detection sensitivity using active infrared thermography. Infrared. Physics and Technology, 68: 52-60.

Lee Y Z, Schubert W. 2008. Determination of the length for tunnel excavation in weak rock.

Tunnelling and Unerground Space Technology, 23: 221-231.

Lemos J V. 1996. Modelling of arch dams on jointed rock foundations. In: Proceedings of ISRM international symposium-EUROCK 96, Turin, Italy: 519-526.

Li S C, Hu C, Li L P, et al. 2013. Bidirectional construction process mechanics for tunnels in dipping layered formation. Tunnelling and Underground Space Technology, 36: 57-65.

Li S J, Yu H, Liu Y X, et al. 2008. Results from in situ monitoring of displacement, bolt load, and disturbed zone of a power house cavern during excavation process. International Journal of Rock Mechanics and Mining Sciences, 45: 1519-1525.

Lin P, Liu H Y, Zhou W Y. 2015. Experimental study on failure behavior of deep tunnels under high in-situ stresses. Tunnelling and Underground Space Technology, 46: 28-45.

Lin P, Ma T H, Liang Z Z, et al. 2014. Failure and overall stability analysis on high arch dam based on DFPA code. Eng. Failure Anal., 45: 164-484.

Liu J, Feng X T, Ding X L, et al. 2003. Stability assessment of the Three-gorges Dam foundation, China using physical and numerical modeling-part I. Physical model tests. International Journal of Rock Mechanics and Mining Sciences, 40(5): 609-631.

Liu Y R, Guan F H, Yang Q, et al. 2013. Geomechanical model test for stability analysis of high arch dam based on small blocks masonry technique. International Journal of Rock Mechanics and Mining Sciences, 61: 231-243.

Lydzba D, Pietruszczak S, Shao J F. 2003. On anisotropy of stratified rocks, homogenization and fabric tensor approach. Computers and Geotechnics, 30: 289-302.

Ma F P, Li Z K, Luo G F. 2004. NIOS model material and its use in geo-mechanical similarity model test. J. Hydraul Electr. Eng., 23(1): 48-51(in Chinese).

Marinos P, Hoek E. 2000. GSI, a geologically friendly tool for rock mass strength estimation. Proceedings of the GeoEng2000 at the International Conference on Geotechnical and Geological Engineering, Melbourne, Australia: 1422-1466.

Marinos V, Marinos P, Hoek E. 2005. The geological strength index, applications and limitations. Bulletin of Engineering Geology and the Environment, 64: 55-56.

Mayr G, Plank B, Sekelja J, et al. 2011. Active thermography as a quantitative method for non-destructive evaluation of porous carbon fiber reinforced polymers. NDT & E Int., 44(7): 537-543.

Mazor D B, Hatzor Y H, Dershowitz W S. 2009. Modeling mechanical layering effects on stability of underground openings in jointed sedimentary rocks. International Journal of Rock Mechanics and Mining Sciences, 46: 262-271.

Mutlu O, Bobet A. 2006. Slip propagation along frictional discontinuities. International Journal of Rock Mechanics and Mining Sciences, 43: 860-876.

Nippon Avionics Co. Ltd. 1990. Instruction manual, Thermal Video System, FINE THERMO, TVS-8000 MKII Series.

Oliverira S, Faria R. 2006. Numerical simulation of collapse scenarios in reduced scale tests of arch dams. Eng. Struct, 28:1430-1439.

Pastor M L, Balandraud X, Grédiac M, et al. 2008. Applying infrared thermagraphy to study the heating of 2024-T3 aluminum specimens under fatigue loading. Infrared Physics and Technology, 51: 505-515

Sagong M, Bobet A. 2002. Coalescence of multiple flaws in a rock-model material in uniaxial compression. International Journal of Rock Mechanics and Mining Sciences, 39(2): 229-241.

Sharma J S, Bolton M D, Boyle R E. 2001. A new technique for simulation of tunnel excavation in a centrifuge. Geotech. Test. J. , 24(4): 343-349.

Shen T, Zou Z S. 1988. A study on geomechanical model materials and exploration of some experimental techniques. J. Yangtze River. Sci. Res. Inst. , 4: 12-23(in Chinese).

Shin J H, Choi Y K, Kwon O Y, et al. 2008. Model testing for pipe-reinforced tunnel heading in a granular soil. Tunnelling and Underground Space Technology, 23(3): 241-250.

Steinberger R, Leitão T I V, Ladstätter E, et al. 2006. Infrared thermographic techniques for non-destructive damage characterization of carbon fiber reinforced polymers during tensile fatigue testing. Int. J. Fatig. , 28: 1340-1347.

Tang Y G, Kung G T C. 2009. Application of nonlinear optimization technique to back analyses of deep excavation. Compt. Geotech. , 36: 276-290.

Tsesarsky M. 2012. Deformation mechanisms and stability analysis of undermined sedimentary rocks in the shallow subsurface. Eng. Geol. , 133-134: 16-29.

Theodorakeas P, Avdelidis N P, Cheilakou E, et al. 2014. Quantitative analysis of plastered mosaics by means of active infrared thermography. Construct. Build Mater. , 73: 417-425.

Tsesarsky M, Hatzor Y H. 2006. Tunnel roof deflection in blocky rock masses as a function of joint spacing and friction-A parametric study using discontinuous deformation analysis (DDA). Tunnelling and Underground Space Technology, 21(1): 29-45.

Wang H P, Li S C, Zhang Q Y, et al. 2006. Development of a new geomechanical similar material. Chinese Journal of Rock Mechanics and Engineering, 25 (9): 1842-1847 (in Chinese).

Zhu W S, Li Y, Li S C, et al. 2011. Quasi-three-dimensional physical model tests on a cavern complex under high in-situ stresses. International Journal of Rock Mechanics and Mining Sciences, 48: 199-209.

Zhu W S, Li Y, Zhang L, et al. 2008. Geomechanical model test on stability of cavern group under high geostress. Chinese Journal of Rock Mechanics and Engineering, 27(7): 1308-1314 (in Chinese).

Zhu W S, Zhang Q B, Zhu H H, et al. 2010. Large-scale geomechanical model testing of an underground cavern group in a true three-dimensional (3-D) stress state. Can. Geotech. J. , 47: 935-946.

Zuo J P, Peng S P, Li Y J, et al. 2009. Investigation of karst collapse based on 3-D seismic technique and DDA method at Xieqiao coal mine. Chin. Int. J. Coal Geol. , 78: 276-287.

Chapter 4   Excavation in 60° inclined strata

4.1 Introduction

Excavation stability connected with energy release has been an active field of research for the community of rock mechanics (Young et al., 2004). The excavation damaged zone (EDZ) around the tunneling face, where in situ rock mass properties and conditions have been altered due to the excavation impact, can be mechanically unstable and could also form a permeable pathway of groundwater flow which would raise the safety concern of the deep geological structures (Zimmerman et al., 2004). The mechanism, degree and extent, therefore, should be better understood and quantified for the creation or operation of an underground opening at depth. This paper presents an experimental study on roadway excavation in 60° inclined stratified rocks based on laboratory experiment and infrared thermography.

The excavation stability related problems have widely been investigated by means of in-situ experiments usually incorporated with numerical methods and acoustic emission (AE) (Young and Collins, 2001; Sitharam and Latha, 2002; Read, 2004; Cai and Kaiser, 2005; Barla, 2008; Tang and Kung, 2009). Mean while laboratory experiments were carried out to investigate the underground excavation problems (Meguid et al., 2008), such as the trap door tests (Tanaka and Sakai, 1993); laboratory experiment using a model specimen (Cheon et al., 2011), and the centrifuge model test (Sharma et al., 2001; Kamata and Mashimo, 2003).

A large number of coal mine in China have thick and steeply inclined coal seams, and mining of them tends to result in geological disasters such as coal bumps, coal and gas outbursts, ground subsidence and floor heave, etc (Ju et al., 2006; Dai, et al., 2013). Failure mechanisms of the steeply inclined rocks, however, have not yet been fully understood due to the problem complexity. One of the major difficulties in the large-scale physical model tests on the steeply inclined strata lies in the monitoring the frictional slip behaviour.

In recent decades as the mining went deeper, investigations on the underground excavation have been focused on the rock mass with discontinuous or weak surfaces, including, for example, numerical simulations (Varadarajan et al.,

2001; Sitharam and Latha, 2002; Jia and Tang, 2008;) and the small-scale model tests (Jeon et al., 2004). Although different degree of successes was achieved in these studies, the obvious insufficiency for the geotechnical research lies in the visualization and observation, in real-time and over the entire field, of the structural effects of the fault-contained and large-scale rock masses.

In order to acquire a better understanding of the excavation-induced damage in stratified rock masses, experiments on roadway excavation using the large-scale geomechanical models and infrared thermography were conducted at the State Key Laboratory for Geomechanics and Deep Underground Engineering (SKL-GDUE), China University of Mining & Technology Beijing (CUMTB), in recent years. This chapter presents the experimental results from the excavation in 60° inclined stratified rocks.

4.2 Experiment

4.2.1 Rock model material

As introduced in section 3.4, the rock strata are constructed with a large number of small prismatic plates, the so-called "elementary slab" which are made of gypsum and water. All the elementary slabs were fabricated with a same surface dimension 400 m×400 m and three thicknesses of 1, 2 and 3 mm, respectively as reported in Table 3.2. A large number of the elementary slabs (shown in Figure 4.1) with the same model-rock property were used to ensemble a rock layer of the same property and a certain number of the rock layers with the same rock property were used to construct a rock stratum.

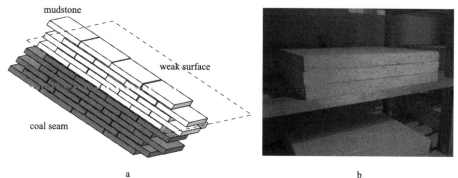

Figure 4.1 Weak bedding surface of the rock strata simulated by the rock-model materials
a. schematic of the rock strata and their interface; b. photograph of the elementary slabs

4.2.2 Geomechanical model construction

The geomechanical model was built with total nine strata including one sandstone, four mudstone and four coal seam; and all the strata inclined at an angle of 60° with respect to the horizontal, to simulate steeply dipped geological structures. The constructed geological model, as shown schematically in Figure 4.2a, is 1.6 m

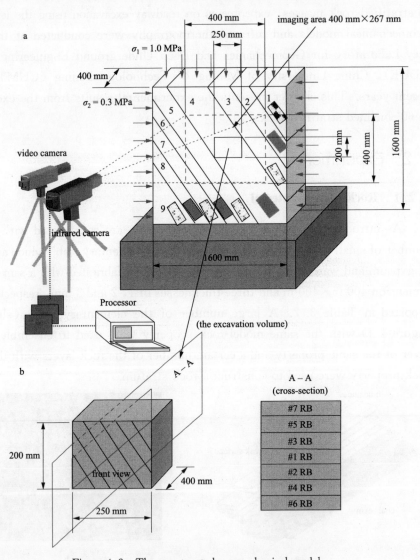

Figure 4.2 The constructed geomechanical model
a. schematic drawing of the geological model; b. the excavation volume (3-dimensional view) and the sectional view (A-A)

high and 1.6 m wide and 0.4 m in thickness. The alternating strata are indexed with 1-9 from the left to the right with the rock properties shown in Figure 4.2a, including the thickness of the strata and the number of layers of the elementary slabs used in each stratum assemblage.

During the excavation, the vertical and lateral loads were applied uniformly on the top and two side boundaries of the model by the loading system stalled in the frame while the bottom of the model was fixed on the basement of testing machine. The lateral pressure vs. vertical pressure σ_2/σ_1 was kept at a constant magnitude of 0.3/1 (MPa) which is equal to the lateral pressure coefficient $\lambda = 0.3$, so as to reproduce the unbalanced stress state found in deep underground mining. The model is able to simulate by the geometric scale factor ($\alpha_l = 12$, see Chapter 3) an engineering rock mass with a plane dimension of 19.2 m×19.2 m and a tunneling face of 3 m×2.4 m.

The excavation zone was located in the center of the model within the stratum 4 (coal seam) and has a dimension of 200 mm in height, 250 mm in width and 400 mm long (equal to the model's thickness). Figure 4.2c shows the total excavation volume (200 m×250 m×400 m) which was divided into seven subspaces termed "rock block (RB)". The RBs were numbered #1—#7 corresponding to the roadway excavation sequence. An overview of the geomechanical model at laboratory was given in Figure 4.3a and the geological section of the model is given in Figure 4.3b.

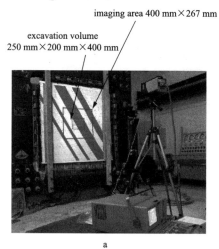

Stratum No.		Rock type	Thickness /mm	Layer number
	1	sandstone	440	14
	2	coal seam	140	10
	3	mudstone	120	7
	4	coal seam	250	17
	5	mudstone	150	5
	6	coal seam	60	4
	7	mudstone	140	5
	8	coal seam	60	3
	9	mudstone	240	8

a   b

Figure 4.3　Photography and geological section of the model
a. photography showing an overview of the model at laboratory; b. geological sections of the model-rock strata

4.2.3 Excavation plan

The roadway excavation was designed as two phases. phase 1 (see Figure 4.4): full-face excavation, i. e., tunneling on #1 RB until a small passage is cut through; and phase 2 (Figure 4.5): staged excavation, i. e., removing one RB at each excavation stage. For description of the full-face excavation, the excavated volume of the artificial materials from the roadway tunnel was referred to as "footage" thereafter and for the staged excavation, the term "excavation stage" denotes the removing of the RB, and total of seven excavation stages were performed step by step in sequence during the phase 2 excavations. The two phase excavation plan is the same as those excavation experimental results conducted in our institute in the published papers (He et al., 2010a, 2010b; He, 2011; Gong et al., 2013).

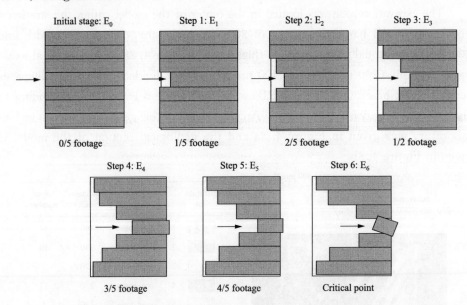

Figure 4.4 The full-face (phase 1) excavation at sectional view

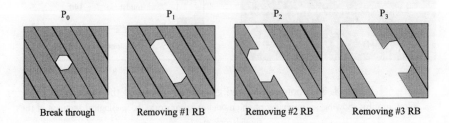

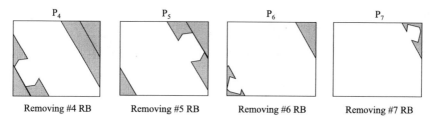

Figure 4.5 The staged (phase 2) excavation at front view

4.2.4 Excavation method

The excavation of a roadway tunnel without support by means of long-round drill and blast was simulated by an operator with a hammer and a chisel. Figure 4.6 shows schematically the scenario where the operator performs the full-face excavation with the hammer and chisel. The excavation was started from the back side of the model and went through to the front face. The infrared camera detects the temperature rise with an imaging area of 400 mm×400 mm.

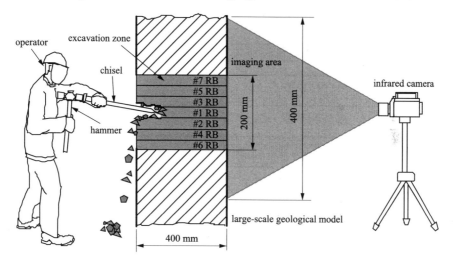

Figure 4.6 Schematic illustration of the phase 1 excavation (full-face excavation)
The tunnel excavation without support with dill and blast method was simulated by an operator
using a hammer and a chisel as the tunneling tools

The hammer-chisel impact feeds an intense and pulsed pressure against the face. The cycles for blowing the chisel with the hammer by the operator are at a slow rate with a random manner. To be exact, the rate of the loading is stochastic, i. e. , when the face is shallow, the blowing was fast (about one blowing every second) as a result of the easy operation; while the face went deeper, the blowing was very slow (around one blow every five seconds or more), because the operator needs to find the right place to be excavated. Thus the excavation-induced impact is, therefore, not periodical and a rate-dependent process but a consecutive-stochastic dynamic loading and unloading process.

Since the excavation area (250 mm×200 mm) is much smaller than the size of the geological model (1600 mm×1600 mm), the vibrations intensity caused by the impact depends on the excavation status. For example, at the beginning of the full-face excavation, the whole model trembled against the blowing at high-frequency with sharp striking sound. When the face advanced deeper, the vibration was localized within the face and the sound was muffled to be a low-frequency noise. Over the stages excavation, the vibration existed in a localized manner. For example, removing of #1 RB caused higher intensity of the vibration than that in removing #7 (the last) RB. The intense vibration occurred at the stage P_1 to stage P_3 excavations. At the last two stages P_6 and P_7, the intensity of the vibrations was minor.

4.3 Infrared detection

4.3.1 Infrared thermography

Infrared (IR) thermography, as a non-destructive, remote sensing technique, has been widely used in detection of deformation and failure process of rock-like materials, based on the fact that the heat generation is caused by the intrinsic dissipation due to elasticity and inelasticity of the material under external loading (Connolly and Copley, 1990; Luong, 1990, 1995, 2007; Steinberger et al., 2006). IR thermography produces heat images directly from the invisible radiant energy(dissipated energy) emitted from stationary or moving objects at any distance.

The widely used non-destructive detection technique includes acoustic emission (AE) and electromagnetic radiation. It describes the failure process of the rock indirectly by transformation or statistics of the test data. On the contrary, infrared thermography directly visualizes the temperature field of the tested object without surface contact or in any way influencing the actual surface temperature of the tested object. Thus the infrared thermographic method can capture the constitutive behavior with more faithful manner.

Generally, infrared thermography is used in two modes, i.e., the passive and the active (Gong et al., 2013). The active thermography uses a known external heat source for heating up the tested object and, therefore, needs to calibrate the obtained temperature increase physically with the heat source. As the heating process is operated in a controlled way, a large temperature increment could be obtained. Thus the temperature-induced stress or strain has larger amplitude and the obtained thermogram has higher contrast, which may contribute to identify the physical process by the subsequent image analysis.

The active infrared thermography is usually used in the case of the materials with a relatively simple constitutive relationship such as metals and composites (Mukhopadhyay and Chanda, 2000; Steinberger et al., 2006). The passive thermography detects the temperature field induced by the external loading based the coupled thermo-mechanical effects without use of any extra heat sources. When using passive thermography, the temperature increment may be relatively smaller and the resultant thermogram is often vague in contrast. Therefore, filtering of the raw thermograms should be implemented (Gong et al., 2013).

TVS-8100 MK Ⅱ infrared thermography was used with the passive mode in this study. The infrared camera operates at wave length of 3.6-4.6 μm, measuring temperature range of 40-300℃, and minimum detection temperature difference of 0.025℃. The infrared camera was placed in the front side of the model at a distance of 1333 mm in order to have an imaging area around 400 mm×400 mm. A red colored frame of plastic thin belt covering an area of 400 mm×400 mm was glued to the model front face centrosymmetric to the roadway cross section, indicating infrared imaging area (Figure 4.1a).

The image acquisition frequency was set as one frame every four seconds. For 24 hours earlier prior to the testing, all the instruments were placed in the same room with the PFESA model, so that the detected IR temperatures are the temperature variation due to the excavation impact. During the test, roadway excavation commenced at the back side of the model while the infrared camera recorded the instant thermal changes at the front face (Figure 4.3). The thermogram sequence were displayed in real-time on the screen in the processor with a 240× 320 pixels and then stored them as 120×160 pixels matrix in the processor.

4.3.2 Thermal-mechanical coupling

Thermal-mechanical coupling effect was investigated for interpretation and quantification of the detected infrared sequence from metals (Luong, 1995;

Pastor et al., 2008) and rock and rock-like materials (Luong, 1990, 2007; Wu et al., 2006a) under laboratory testing conditions. Based on the second law of thermodynamics and compatibility and separability of the fundamental equations of mechanics and constitutive relations, the thermo-viso-elastic-plasticity equations were developed by Luong (1995).

The final results of the coupled thermomechanical equation take a universal form (Luong, 1995),

$$\rho C_v \dot{T} = \rho r + \text{div}(K \text{grad} T) - (\beta : \overset{4}{D} : \dot{E}^e) T + S : \dot{E}^I \qquad (4.1)$$

where, ρ is the mass unit in the reference solid, kg/m^3; C_v is the specific heat at constant deformation, $J/(kg \cdot K)$; T is the absolute temperature and the superposed dot stands for the material time derivative; r is the heat supply; div is the divergence operator; K is the thermal conductivity, $W/(m \cdot K)$; grad is the gradient operator; β is the coefficient of the thermal expansion matrix; ":" is the contracted product operator of second order tensor; $\overset{4}{D}$ is the fourth-order elasticity tensor; E^e is the elastic strain tensor; E^I is the inelastic strain tensor; S is the second Piola-Kirchhoff stress tensor; The volumetric heat capacity of the material $C = \rho C_v$ is the energy required to raise the temperature of unit volume by 1℃ (or 1 K).

This equation delineates the four different thermal-mechanical coupling mechanisms, represented by the four terms on the right-hand side of the Eq. (4.1) underlying rock failure processes (Luong, 1995), including sources or sinks of hear in the scanning spot (the first term), thermal conduction (the second term) related to the transference of heat by thermal conduction to rebalance the temperature distribution of the stressed object, thermoelasticity (the third term) representing the thermal-elastic coupling effect which is a linear process and the conversion between the mechanical and thermal energy is reversible, and energy dissipation (the last term) generated by viscosity and/or plasticity involving expanding of initial fissures and joints, generating of new fractures, fracture propagation and extension, pore gas desorbing-escaping, and rock friction, etc. It is a non-linear process and energy conversion is irreversible.

Detailed explanation of Eq. (4.1) can be found in the references (Luong, 1995, 2007). Although the principles was made clear, but the discrimination of the four quite different phenomena is difficult, especially for the intrinsic dissipation, when interpreting the thermograms obtained under usual conditions as pointed out by Luong (1995). In this study, image denoising and data reduction were used to ascertain the four different mechanisms of the thermal-mechanical coupling effect over the excavation processes.

4.4 Image processing

4.4.1 Problem statement

Real world rarely comes clean. Noise contamination problems are commonly existed in the graphical testing techniques. Filtering the noisy image while preserving the image features such as edges and textures is of central importance for making progress in the infrared thermography techniques (Luong, 1995). At laboratory infrared thermography is often used to detect small-scale objects, e. g. the laboratory-sized specimen (Wu et al., 2002, 2004, 2006a, 2006b) or used in the active mode (Luong, 1995; Steinberger et al., 2006; Pastor et al., 2008).

In this case, the image resolution is higher due to the small imaging region (defined as "pixel/ unit area") or the high radiant flux simulated by the extra heating source. Thus the requirements for the image processing may be lower. However, in the case of detecting a large-scale geological model as in our test with the infrared thermography used in the passive mode, the thermograms are highly contaminated by the noises with a narrowed dynamical range. Therefore, for the low-quality thermograms, image processing procedure should be properly performed.

4.4.2 Algorithms

1. The types of noise

Processing of a low-quality thermogram often involves such procedures as image denoising and enhancement (Gong et al., 2013). Major task for the image denoising is the estimation of the noises that may exist in a specific physical process and the recovery of the clean image by the use of the appropriate filters. For laboratory mechanical experiments, the noises contained in the thermogram can be divided into the following classes, i. e. , the environmental radiation noise, electronic current induced impulsive noise (also salt-and pepper noise) and additive-periodical noises.

2. Image subtraction

The environmental radiation noises can be removed by image subtraction. The image subtraction algorithm can be given by,

$$f_k(x,y) = f_k(x,y) - f_0(x,y) \qquad (4.2)$$

where, k is the frame index of the infrared sequence; $f_0(x,y)$ is the first frame of the infrared sequence taken when the model rock was at intact state; $x = 0,1,2,\cdots,N-1$, $y = 0,1,2,\cdots,M-1$ are the pixel coordinates, and $M=120$ and $N=160$ are the maximum pixel number respectively for the image matrix. Subtraction of the first frame from the following infrared sequence will suppress the background radiation noises.

3. The median filter

The existence of the impulsive noises can narrow the dynamical scope and make the following image processing improbable such as the image enhancement. The impulsive noises can be removed by the *median* filter (Pratt, 1991; Gonzalez et al., 2005). The *median* filter is a non-linear spatial filter or rank filter whose response is based on ordering the pixels contained in the image area encompassed by the filter, and replacing the central pixel with the ranking result.

Given S_{xy} representing the structural element (or mask) centered at the point (x, y) which actually is a matrix of $m \times n$ dimension, $g(x, y)$ representing the filtered image, and $\hat{g}(x,y)$ the noisy image, then the *median* operation is replacing the pixel at (x, y) with the median pixel of the neighborhood defined by the mask,

$$g(x,y) = \underset{(s,t) \in S_{xy}}{\operatorname{median}} \{\hat{g}(s,t)\} \qquad (4.3)$$

Performing the median filtration on the thermogram can get rid of the pixel with much larger magnitude in its neighborhood area and enlarge the dynamical scope of the image. It is a necessary step for the subsequent image enhancement operation.

The 2D median filtering was computed by the standard filtering function "*medfilt2*" in the IPT with a 4-neighbor 3×3 mask at the first step. The median filtering operation will smooth the image by removing some singular values of the gray level matrix. In essence, the presence of the singularity in the pixel matrix caused by the pulse noise narrows the scale of the magnitude for the real information represented by the pixels; and the removing the pulse noise will amplify the scale of the magnitude of the pixels and make it possible to exhibiting the details of the thermal images.

Assuming that the noises left in this step is only the additive-periodical one after image subtraction and *median* filtering, the additive noise-contained image $g(x, y)$ can be represented by,

$$g(x,y) = f(x,y) + \eta(x,y) \qquad (4.4)$$

where, $f(x, y)$ is the clean image; $\eta(x, y)$ is the left noise; and (x, y) are the horizontal and vertical coordinates of the image plane respectively and take the discrete value in the case of digital image. For reduction of the additive-periodical noise, frequency-domain methods can be used and Gaussian-high-pass filter (GHPF) is proved to be efficacious in implementation of this task (Gonzalez et al., 2005).

4. The Gaussian-high-pass filter

Implementation of the Gaussian-high-pass filter involves the following four steps.

(1) Firstly compute Fourier transform of Eq. (4.4), then yields,

$$G(u,v) = F(u,v) + N(u,v) \qquad (4.5)$$

where, for digital image, 2-dimensional Discrete Fourier Transform (DFT) was employed and $G(u,v) = \text{DFT}[g(x,y)]$, $F(u,v) = \text{DFT}[f(x,y)]$, $N(u,v) = \text{DFT}[\eta(x,y)]$, DFT is the discrete Fourier transform operator (Gonzalez et al., 2005); u and v are the frequency variables in the horizontal and vertical directions, Hz.

(2) Let $H(u,v)$ stand for Fourier transform of the filtering function, multiply Eq. (4.5) with $H(u,v)$ and assuming that the product of $H(u,v)N(u,v)$ vanishes, the filtering algorithm can be implemented by

$$G(u,v) = H(u,v)F(u,v) \qquad (4.6)$$

(3) Perform the inverse Fourier transformation on $G(u,v)$ in Eq. (4.6), i.e., $g(x,y) = \text{DFT}^{-1}[G(u,v)]$, then the cleaned image can be recovered from $g(x,y)$. The transfer function $H(u,v)$ used in this research is the GHPF (Gaussian-high-pass filter) which can be mathematically represented by (Gonzalez et al., 2005),

$$H(u,v) = 1 - e^{-D^2(u,v)/2D_0} \qquad (4.7)$$

where, D_0 is the cut off frequency of the filter; and D is the distance from a point (u, v) to $(M/2, N/2)$ which is the center for an $M \times N$ image,

$$D(u,v) = [(u-M/2)^2 + (v-N/2)^2]^{1/2} \qquad (4.8)$$

Figure 4.7 shows a three-dimensional representation of the GHPF. GHPF can filtrate the low-pass band components (located around center of the image plane, the horizontal plane in the figure) while preserve high-pass band components. Edges and borders that segment different objects in an image are important image features corresponding to the high-frequency components. Compared with other frequency-domain high-pass filters such as the ideal high-pass filter and butterworth high-pass filter, GHPF can preserve fine features with smoother visual effect (Gong et al., 2012).

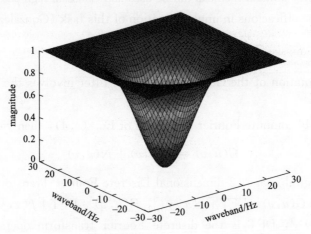

Figure 4.7 Three-dimensional representation of Gaussian-high-pass filter (GHPF)

4.4.3 Processing and assessment

The algorithms in the previous sub-section were realized using the MATLAB 8.0 macro code and functions in IPT (Image Processing Toolbox in MATLAB). Figure 4.8 shows the image processing effect. Figure 4.8a is the input thermogram of frame 587 corresponding to E_6 in the full-face excavation. The raw thermogram is a fully-blurred low contrast image with narrow dynamic range as a result of the contamination by the noises. No any specific and clear information could be draw from it. Figure 4.8b is the processed thermogram of the frame 587 using the proposed algorithm described above.

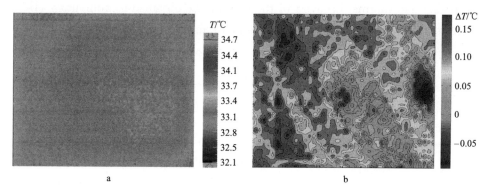

Figure 4.8 Taking the frame 587 in the sequence corresponding to E6 in the full-face excavation being a case, illustrating the image processing effect using the algorithm proposed in this paper
a. the raw thermogram; b. the processed one

As compared to the raw thermogram (see Figure 4.8a), the borders and edges of the excavation-induced frictional sliding bands with changed stress represented by the changed brightness can be distinguished more clearly in the processed thermogram (see Figure 4.8b). The image processing algorithm proposed in this paper, therefore, can be said very efficacious in noise reduction and image enhancement. The thermogram analysis method, taking the processed thermogram of frame 587 as an example, will be given in the next section.

4.5 Image analysis

4.5.1 Extracting the energy release index

The thermography matrix data sets is the infrared radiation temperature (IRT) distribution on the viewed surface, which is essentially the rock responses as a result of the thermomechanical coupling nature of the engineering materials. Let $f(x,y)$, x and y are the pixel coordinates ($x=1,2,\cdots,M$; $y=1,2,\cdots,N$), denote the thermal image matrix data at an instant time, where M and N are the maximum pixel numbers ($M=120$ and $N=160$).

Image matrix $f(x,y)$ is actually the surface temperature field of the tested object. The mean of the image matrix $\langle f(x,y) \rangle$, also represented by $\langle IRT \rangle$ in the following context, accounts for the averaged energy dissipation at a time instant. Thus, $\langle IRT \rangle$ can be viewed as the overall thermal rock response to the external load and be defined as the *energy release index*. Thus evolution of the comprehensive thermal behavior of the rock can be obtained by computing ⟨IRT⟩ from the infrared sequence using the standard algorithm.

The *energy release index* ⟨IRT⟩ can be obtained by computing the statistical mean on the thermal image matrix,

$$\langle \text{IRT} \rangle = \frac{1}{M}\frac{1}{N}\sum_{x=1}^{N}\sum_{y=1}^{M} f(x,y) \qquad (4.9)$$

The averaged IRT distribution, ⟨IRT⟩, is actually the density of the infrared emittance at a time instant. By computing ⟨IRT⟩ on the infrared sequence, a time series, $\langle \text{IRT} \rangle_k$, is obtained, $k = 0, 1, 2, \cdots$, where k corresponds to the sampling time.

The first frame of the infrared image was taken when the model was at the intact state, which was used as a benchmark reference for calibration of the temperature changes of the following IR images. We used a normalized $\langle \text{IRT} \rangle_k$ as a measure in characterizing rock behaviors over the roadway excavation. For the sake of simplicity, the normalized $\langle \text{IRT} \rangle_k$ is also referred to as "IRT", which actually represents: ①IRT variations relative to the model's intact state, and ②the energy release level relative to the maximum mean value of IRT.

4.5.2 Spectral characterization

1. Fourier transform

Taking the denoised thermogram of the frame 587 as an example, our image analysis method is shown in Figure 4.9. The image characterization include: ①the denoised thermogram itself (Figure 4.9c); ②mean value of the infrared temperature (IRT) field (Figure 4.9b) and ③spatial Fourier spectra of the image (Figure 4.9c,d).

Conventionally for an image, two-dimensional Fourier transform is generally used. However, it is inconvenient for a quantitative analysis of spectral features. Alternatively, one-dimensional (1-D) Fourier transform can be used for computing the spectra of the re-sampled 1-D "spatial series". As shown in Figure 4.9c, two spatial series, $x_i (i=1,2,\cdots, 120)$ and $x_j (j=1,2,\cdots, 160)$ were sampled along the horizontal and vertical axes across the center of the image plane.

2. Wave number and wave length

Compute 120 and 160 points discrete Fourier transform (DFT) on the two

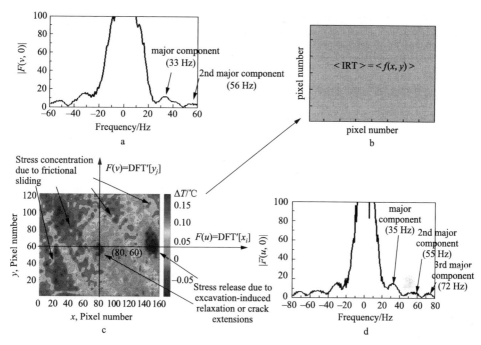

Figure 4.9 Taking the denoised thermogram of the frame 587 corresponding to E_6 as an example, the image analysis method used in our study is illustrated here
a. the denoised thermogram; b. the mean value of the image matrix ⟨IRT⟩; c. the vertical spatial spectrum $F(v)$ obtained by 120 points DFT; d. the horizontal spectrum $F(u)$ obtained by 160 points DFT

data sets respectively, then we have horizontal spatial spectrum $F(u)=\text{DFT}[x_i]$ and vertical spatial spectrum $F(v)=\text{DFT}[y_j]$; u (Hz) and v (Hz) are the horizontal and vertical spatial frequencies (or wave number) respectively and take discrete values. Variable $k=\sqrt{u^2+v^2}$ is the spatial frequency (or wave number) of arbitrary direction and k is connected with wave length λ by the relation $\lambda \propto 1/k$ (Gong et al., 2008). Hence, propagation features of the stress wave can be analyzed from the two spatial spectra.

The wavelength parameter in a dynamical process has an significant effect on rock damage (Brady and Brown, 2004). In principle, the dynamical stress induced damage on a tunnel of characteristic length D in the path of a plane seismic wave with wave length λ is related to their relative magnitude. If the size of the tunnel is negligible as compared to the wave length, i.e., $D\lambda^{-1}$ approaches zero, the stress wave induced loading on the tunnel is similar to the static loading; if the tunnel size is infinite as compared to the wave length, i.e., $D\lambda^{-1}$ approaches to infinity, the theory of wave reflection of the propagating waves applies.

In this case, the adjacent rock mass will experience dynamical impulsive stress with respect to the high frequency components. In a spatial spectrum, therefore, high frequency component is the failure precursors. From Figure 4.9c and d, the horizontal spectrum has higher frequency components as compared to the vertical spectrum, indicating that degree of damage is higher in the horizontal direction than that in the vertical. The Fourier spectrum is complete symmetric in a mathematical sense. So either positive or negative half plane can be used to perform the spectral analysis. Some minor distortions in the spectrum plots are due to computational digital error.

4.5.3 Principles for image analysis

Comprehension and interpretation of an thermogram rely on such image features as colors, borders and edges that segment different zones in the image, as well as the temperature scale. The temperature scale represents the incremental temperature relative to the background radiation as a result of the subtraction of the first frame in infrared sequence as described in section 4.4. In an infrared image, pseudo-colors are generally used to highlight the temperature levels. Hot (or positive) colors stand for high-level temperatures and cool (or negative) colors for the low-level temperatures.

A great number of studies demonstrated that higher temperature corresponds to higher stress level due to friction, shearing or stress concentration etc., and lower tepmerature corresponds to lower stress level due to tensile cracking, stress release or unloading, etc., (Connolly and Copley, 1990; Luong, 1990, 1995, 2007; Grinzato et al., 2004; Steinberger et al., 2006; Wu et al., 2006a; Pastor et al., 2008). Besides, the borders or edges for hot and cool colored areas represent different modes of rock behaviors. For example, in the denoised thermogram (Figure 4.9a), the red-colored strips formed by the scattered IRT distribution indicates the frictional effect; and the localized cool-colored areas represent plastic damage due to sudden expansion of the macroscopic cracks. (For comprehending the meanings denoted by colors, please see the Appendix at the end of this book)

4.6 Experimental results

4.6.1 Overall thermal response

Figure 4.10 shows the evolution of ⟨IRT⟩ against face development with

Chapter 4 Excavation in 60° inclined strata

respect to the time sequence. The capital letters E_0-E_6 standing for the excavation steps over the full-face excavation (also see Figure 4.1) and P_0-P_7 standing for the excavation stages during the staged excavation (also see Figure 4.2). As mentioned above, the averaged infrared temperature field ⟨IRT⟩ represents the comprehensive energy dissipation level (overall thermal response to the excavation) of the model rocks. During the full-face excavation, ⟨IRT⟩ curve can be approximated as linear elastic increase over E_0-E_3; the sharp drop of ⟨IRT⟩ immediately after the step 3 excavation represents a frictional sliding event caused by the excavation; and after the IRT drop, ⟨IRT⟩ resume the linear-elastic increase until E_6. During the staged excavation, ⟨IRT⟩ curve develops nonlinearly. At the excavation stages P_0-P_3, energy dissipated at relatively higher level. From the ⟨IRT⟩ curve we see that removing #2 RB results in the highest energy dissipation level and unloading effect represented by the immediately sharp drop of ⟨IRT⟩. Removing #3 RB also caused higher level of ⟨IRT⟩ and a significant ⟨IRT⟩ drop. Thus #2 and #3 rock blocks can be regarded as the rock support in the excavation zone. At the excavation stages P_4-P_7, both the temperature level and the ⟨IRT⟩ drops are smaller, indicating the fact that these RBs have less effect on the excavation stability. Detail rock responses will be given by interpretation of the denoised thermograms in the following context.

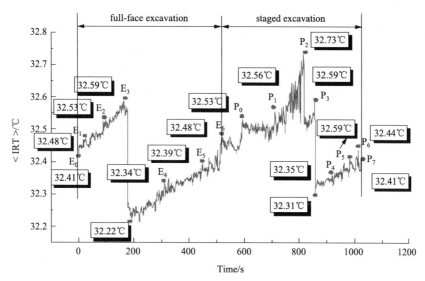

Figure 4.10 Evolution of ⟨IRT⟩ against face development with respect to the time sequence

4.6.2 Heat sources and thermal conduction

The coupled thermomechanical Eq. (4.1) provides a consistent theoretical framework for interpretation and comprehension of the obtained thermograms. As pointed out by Luong (1995; 2007), the detected temperature change resulting from quite different phenomena represented by the four terms at the right hand side of the Eq. (4.1), i.e., the heat sources, the thermal conduction effect, the reversible thermo-elastic coupling and intrinsic dissipation, must be correctly discriminated by such means as particular test conditions and/or specific data reduction which is the main difficulty when interpreting the thermal images obtained from experiments under the usual conditions (Luong, 2007).

The first term on the right hand side of Eq. (4.1) represents the existence of hear sources. The heat sources in our test come from the loading platen and the excavation method. Since the imaging zone is far away from the loading platen and boundary load was kept constant during the excavation, the generated undesirable heat by the loading platen that may obscure the intrinsic dissipation is minor. As for the excavation method (see sub-section 4.2.2 and Figure 4.3), the excavation tool will affect the infrared radiation temperature field at some specific excavation status, which will be analyzed later.

The volumetric heat capacity of the material $C = \rho C_v$ is the energy required to raise the temperature of a unit volume by 1℃ (or K). The second term governs the transference of heat by thermal conduction in which the heat passes through the material to make the temperature uniform and ratio of the thermal conductivity to the heat capacity $\alpha = k/C$, termed the thermal diffusivity, will become the governing parameter in the dynamical process (Luong, 1995). Considering that both the thermal conductivity and diffusivity for the rock materials is smaller than metals (Luong, 2007) and excavation impact is transient, the heat conduction influence is not global but only in the localized and transient manner.

4.6.3 Characterization of the full-face excavation

For discrimination of the four different heat-generation mechanisms in the coupled thermomechanical equation, differentiating the thermo-elastic coupling and intrinsic dissipation are of central importance in comprehension and interpretation of an infrared image. Although various test conditions were developed in the past decades, a marked advancement in this aspect may be difficult especially in the community of RSRM (Remote Sensing Rock Mechanics). In this study as suggested by Luong (1995), the imaging processing based method was used for the differentiation of the thermoelasticity and thermoplasticity effects in the denoised thermograms.

Figure 4.11-Figure 4.17 shows the figure sets for the excavation steps from E_0 to E_6. In each of the figure set, the first figure (indicated by letter a) is the excavation diagram representing the drivage over the full-phase excavation; the second figure is the corresponding denoised thermogram (indicated by b), and the last two figures (marked by c and d) are the horizontal and the vertical Fourier spectra respectively.

Figure 4.11 is the figure set for the initial state of the excavation when the model was not undergone the excavation impact. Figure 4.12 corresponds to the step 1 excavation. By comparison between the two thermal images (Figure 4.12a,b), it is seen that that the IRT, i. e. , the hot-colored and cool-colored areas, distribute with a scattering-random manner indicating the elastic response of the model rocks. The components in the horizontal and vertical spectra distributes with a narrow band and very small amplitude. The thermoelastic response can be identified by these thermal and spectral features. The temperature change within elastic range in a stressed solid is known as thermo-elastic coupling effect (Luong, 2007) which is manifested by the third term on the right hand side of Eq. (4-1). Within the elastic range, a material experiences a reversible conversion between mechanical and thermal energy causing it to change temperature under the condition of adiabatic conditions.

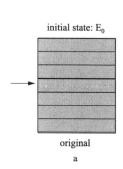

a
original
initial state: E_0

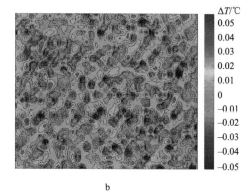

b

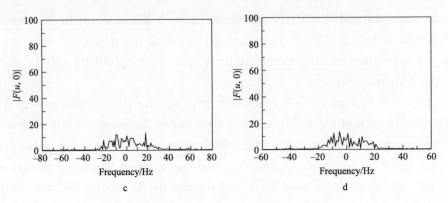

Figure 4.11 Figure set for the state E_0

a. intact state of the model; b. denoised thermograms; c. the horizontal spectrum $|F(u, 0)|$;
d. the vertical spectrum $|F(0, v)|$

Given that the material is isotropic and linear elastic, the integral representation of the thermoelastic term is given by Reference (Wu et al., 2006b) for a plane stress state:

$$\Delta T = -\frac{\beta}{\rho C_v} T \Delta(\sigma_1 + \sigma_2) \quad (4.10)$$

where, T is the surface absolute temperature of loaded solid, K, ΔT is the Temperature increment, K, β is the factor of linear expansion, K^{-1}; ρ is the solid density, kg/m^3; C_v is the thermal capacity of solid at normal atmosphere, $J/(kg \cdot K)$; σ_1 and σ_2 are the two principal stresses of rock surface, MPa. As analyzed above, the IRT distribution pattern depicted in thermograms E_0 (Figure 4.11) and E_1 (Figure 4.12), i.e., the scattering-random distribution of IRT, can be viewed as a typical manifestation of the thermoelastic behavior of the rock.

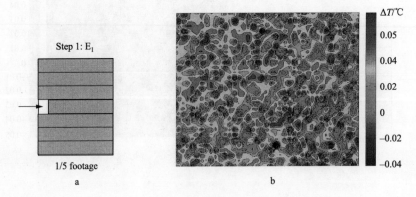

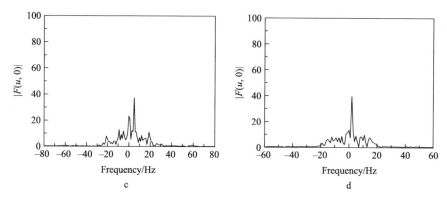

Figure 4.12 Figure set for the excavation step E_1
a. step 1 excavation; b. denoised thermograms; c. the horizontal spectrum $|F(u, 0)|$;
d. the vertical spectrum $|F(0, v)|$

As illustrated in Figure 4.1, the rock layer was constructed with elemental slabs and rock stratum was constructed with rock layers. Excavation impact will cause static friction between the slabs, rock layers and rock strata. Thus the friction is a multi-scale phenomenon like in the loaded rock; friction action exists between mineral molecules, grains, joints, fissures and fractures (Wu et al., 2006b; Luong, 2007). Although ⟨IRT⟩ curve (Figure 4.10) indicates the thermoelasticity for the comprehensive thermal response with respect to E_0-E_3 and E_4-E_6, the denoised thermograms present the detailed structural information. That is, thermograms E_0 and E_1 represent the pure thermoelasticity manifested by the scattering-random IRT. Thermograms E_2 (Figure 4.13b) represents the thermo-elastic behavior with weak static friction indicated by scattered IRT directionally distributed as stratum-paralleled strips with low contrast; and thermograms E_3-E_6 (Figure 4.14b-Figure 4.17b) represent the thermo-elastic behavior with strong static friction illustrated by the higher contrast for the IRT strips.

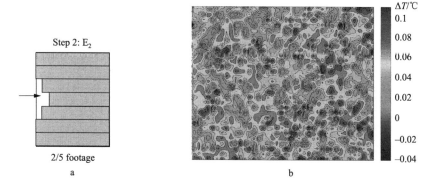

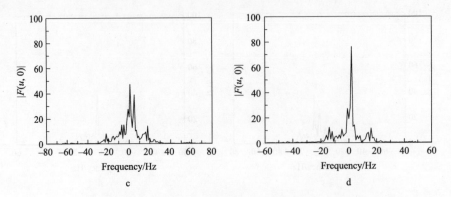

Figure 4.13 Figure set for the excavation step E_2
a. step 2 excavation; b. denoised thermograms; c. the horizontal spectrum $|F(u, 0)|$;
d. the vertical spectrum $|F(0, v)|$

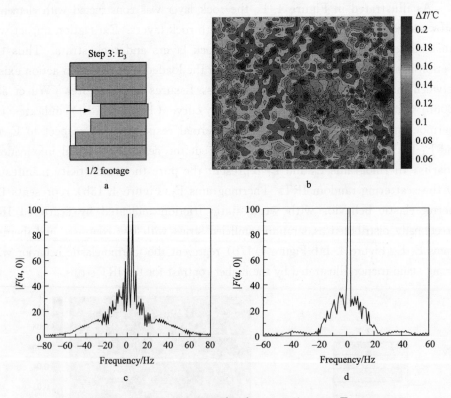

Figure 4.14 Figure set for the excavation step E_3
a. step 3 excavation; b. denoised thermograms; c. the horizontal spectrum $|F(u, 0)|$;
d. the vertical spectrum $|F(0, v)|$

Chapter 4　Excavation in 60° inclined strata　　　　　　　　　　　　　　　• 101 •

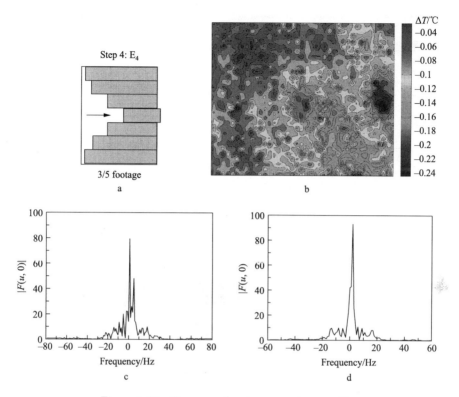

Figure 4.15　Figure set for the excavation step E_4
a. step 4 excavation; b. denoised thermograms; c. the horizontal spectrum $|F(u, 0)|$;
d. the vertical spectrum $|F(0, v)|$

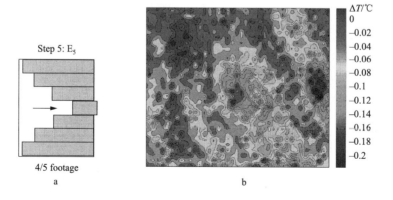

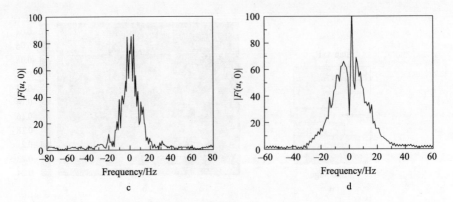

Figure 4.16 Figure set for the excavation step E_5
a. step 5 excavation; b. denoised thermograms; c. the horizontal spectrum $|F(u, 0)|$;
d. the vertical spectrum $|F(0, v)|$

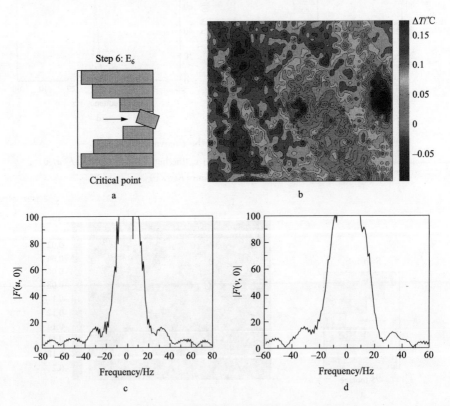

Figure 4.17 Figure set for the excavation step E_6
a. step 6 excavation; b. denoised thermograms; c. the horizontal spectrum $|F(u, 0)|$;
d. the vertical spectrum $|F(0, v)|$

These findings demonstrate the heterogeneity and anisotropy of the steep-inclined stratified rocks. At the same time, the microscopic damage induced by the static friction took places at the very early excavation phase where no macroscopic displacement of the surrounding rock mass occurred since the excavation excavation-induced tangential force not exceeding the critical resistance force defined by the Amonton's law (Carpinteri and Paggi, 2005).

Fourier spectra also characterize the degree of the excavation-induced damage by the energy dissipation level represented by the amplitude of zero-frequency component (also known as "direct-current component") and major wave component (i. e. , the distinct peaks in the mid to high frequency bands). Higher amplitude for the direct-current component indicates higher-level of the thermal energy produced by the thermal-mechanical coupling effect. More wave components in the higher frequency bands denote the severe excavation damage.

4.6.4 Heat production mechanism in the staged excavation

1. Comparison between the energy release between the full-face excavation and the stated excavation

For better understanding of thermal responses during the staged excavation, it is necessary to go back to section 4. 10 to talk further about the ⟨IRT⟩ curve in Figure 4. 10. During the full-face excavation, tunneling was confined within the #1 RB with a small face area and there is no macroscopic displacement occurred in the adjacent rocks. While during the staged excavation, the face was expanded in size as the removing of each of the RBs which caused the macroscopic displacements in the country rocks.

This is why the time-marching scheme for ⟨IRT⟩ is multi-linear over the full-face excavation and non-linear over the staged excavation. The multi-linearity and non-linearity of the ⟨IRT⟩ profile are divided by point E_6 at which #1 RB was destructed for the first time. Note that in the most cases ⟨IRT⟩ level over the full-face excavation is higher than that over the staged excavation. It is due primarily to the combined two factors including the intrinsic dissipation and the influence by the manned tunneling operation, and interpretation of which will be given in what follows.

2. Heat production

The last term on the right hand side of the coupled thermomechanical Eq. (4.1) manifests the intrinsic dissipation generated by viscosity and/or plasticity (Luong, 1995, 2007). The heat production mechanism of internal energy dissipation caused by plastic deformation was experimentally identified by research works such as in the references (Luong, 1990, 1995, 2007; Wu et al., 2002, 2004, 2006a, 2006b; He et al., 2010a, 2010b; He, 2011; Gong et al., 2013). It is generally accepted that in thermo-elastic-plasticity, not all the mechanical work produced by plastic deformation can be converted to the thermal energy in a solid, i. e. , a portion of the work is spent in the change of material microscopic structure (Luong, 1990).

The plastic deformation related dissipation involves fracturing and friction in our test. As production of new fractures, extension of the existed fracture as mode I crack need to consume energy, thus the temperature change for this part of the dissipation is always negative depicted in the thermogram as cool colors. Friction between grains, joints and fractures of different scales as mode II cracks constitutes a principal part of the plastic deformation in the stressed rock. The friction-produced heat is always positive and depicted in the thermogram as hot colors.

4.6.5 Characterization of the staged excavation

Figure 4.18-Figure 4.25 show the figure sets for the excavation stages from P_0 to P_7. In each of the figure set, the first figure (indicated by letter a) is the excavation diagram representing the drivage over the full-phase excavation; the second figure is the corresponding denoised thermogram (indicated by b), and the last two figures (marked by c and d) are the horizontal and the vertical Fourier spectra respectively.

Detailed heat patterns of the plastic damage can be seen in the denoised thermograms shown in Figure 4.18-Figure 4.25. Thermal image P_0 (Figure 4.18b) depicts the instant when a small passage was firstly cut through in #1 RB. For drilling a hole through, the operator needs to blow the chisel at his best at the back side of the model until the chisel went through to the front side which thus as a intense heat source detected by the infrared camera. This strong heat source narrows the dynamical scope of the surface temperature field.

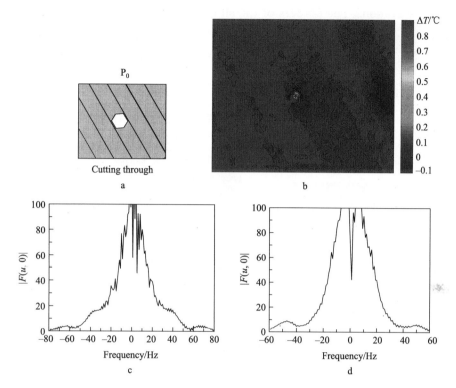

Figure 4.18 Figure set for the excavation stage P_0 (colored image presented in the Appendix) a. cutting through a small passage in #1 RB; b. denoised thermograms; c. the horizontal spectrum $|F(u, 0)|$; d. the vertical spectrum $|F(0, v)|$

In the thermal image P_0, therefore, the cutting-through area was seen as a red-colored dot and elsewhere is deep blue colored. However, there are still deeper-blue colored IRT strips seen in the image P_0 (Figure 4.18b) indicating the frictional effect. The direct-current wave component and high-frequency components have high amplitude accounting for high dissipation level and degree of the damage (see Figure 4.18c, d).

The plastic deformation, whereby the excavation-induced frictional sliding causing permanent changes globally or locally, is one of the most effective heat-production mechanisms. Thermal images P_1 and P_2 (Figures 4.19b and 4.20b) depict the thermal response with respect to removing of #1 and #2 RBs. Higher level of the energy dissipation (the highest for P_2) is the common features for the two thermograms accounting for the higher degree of the plastic damage caused by the excavation. The borders of the slip bands are not so clear, which denotes that the excavation-induced tangential force did not attain the critical value of the static frictional resistance defined by the Amonton's law. Note that there is a minor dynamic sliding immediately after the excavation stage 2 (see Figure 4.10).

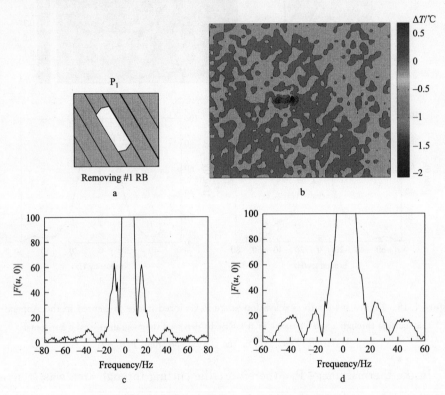

Figure 4.19 Figure set for the excavation stage P_1
a. staged 1 excavation (removing of #1 RB); b. denoised thermograms; c. the horizontal spectrum $|F(u, 0)|$; d. the vertical spectrum $|F(0, v)|$

It is seen from the vertical Fourier spectrum (see Figure 4.19d) that high frequency band components were existed with relatively high amplitudes, indicating the fact that removing the #1 rock block caused intense stress waves propagating along the vertical direction which is a sign for occurrence of the excavation induced damage in that direction. In contrast, in the horizontal Fourier spectrum, the high-band components were existed with smaller amplitude indicating the less intense wave propagation in the horizontal direction.

By observing the horizontal spectrum (Figure 4. 20b) and vertical spectrum (Figure 4. 20c) for the excavation stage P_2, one can see that the high-band components were existed both in the horizontal and vertical spectra with high amplitude. The amplitude of the high-band components are higher than that of the spectra for the excavation stage P_1, forecasting the heavier excavation damage.

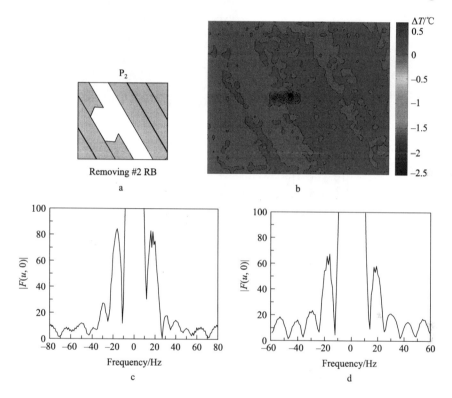

Figure 4. 20　Figure set for the excavation stage P_2

a. staged 2 excavation (removing of #2 RB); b. denoised thermograms; c. the horizontal spectrum $|F(u, 0)|$; d. the vertical spectrum $|F(0, v)|$

In contrast, in thermal image P_3 (Figure 4. 21b), the borders of the slip bands are very sharp indicating the dynamic friction between rock layers defined by the Coulomb's law. In accordance with the removing of #3 RB, the excavated face was expanded so large as to the relative sliding of the rock layers was initiated, which caused a significant drop in the thermal temperature as illustrated in Figure 4. 10. The high amplitude wave components distribute in the broader band accounting for the high-degree excavation damage induced by removing of #3

RBs (see Figures 4.21c,d). Therefore, the #3 rock block should be referred to as the "critical RBs".

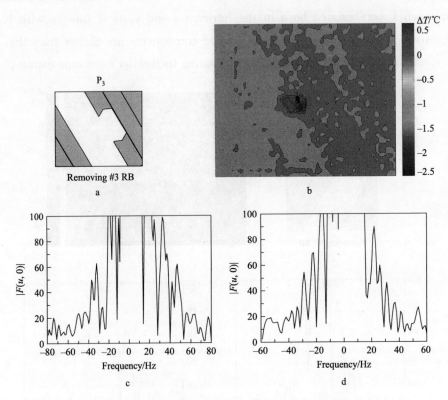

Figure 4.21 Figure set for the excavation stage P_3
a. staged 3 excavation (removing of #3 RB); b. denoised thermograms; c. the horizontal spectrum $|F(u, 0)|$; d. the vertical spectrum $|F(0, v)|$

In thermal image P_4 (see Figure 4.22b), the strip-formed IRT distribution was seen. The borders between the hot-colored and cool-colored IRT strips are clear, indicating the dynamical frictional slipping effect of the rock layers. From the thermal patterns of the analyzed thermograms, it is clear that with regard to the damage mechanism for the steeply dipped rock strata, the slip-induced plastic deformation is dominant. In thermal images P_5 and P_6 (see Figures 4.23b and 4.24b), the borders of the slip bands are not very clear and the IRT distribution tends to become even, indicating the less degree of the excavation-induced damage. In thermal image P_7 (see Figure 4.25b), the IRT distribution come back to the scattering-random pattern indicating that fact that the thermal-mechanical coupling effect for removing of #7 RB (the last RB) in minor and the stress redistribution in the surrounding rocks was over. From the Fourier spectra from the excavation stages P_4 (see Figures 4. 22c、d and 4. 25c、d), the high-band components both in the horizontal and vertical spectra have smaller amplitude as compared to the previous stages, indicating the fact that excavation of the rock blocks from #4 to #6 did not cause sewer rock damage in comparison to the previous excavation stages. Therefore, degree of the excavation-induced damage can also be identified according to the wave propagation patterns such as the amplitude and frequency bands.

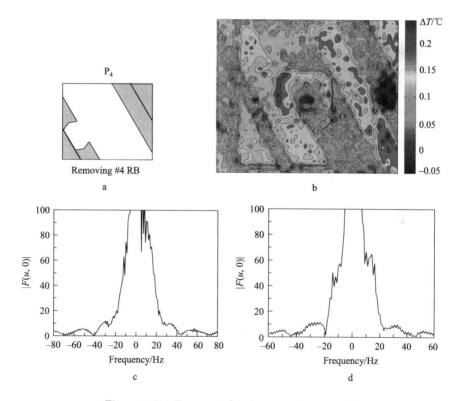

Figure 4. 22　Figure set for the excavation stage P_4

a. staged 4 excavation (removing of #4 RB); b. denoised thermograms; c. the horizontal spectrum $|F(u, 0)|$; d. the vertical spectrum $|F(0, v)|$

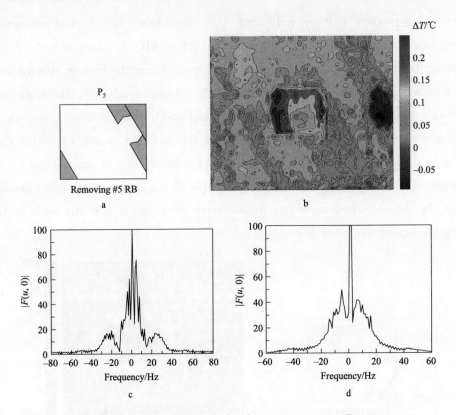

Figure 4.23 Figure set for the excavation stage P_5

a. staged 5 excavation (removing of #5 RB); b. denoised thermograms; c. the horizontal spectrum $|F(u, 0)|$; d. the vertical spectrum $|F(0, v)|$

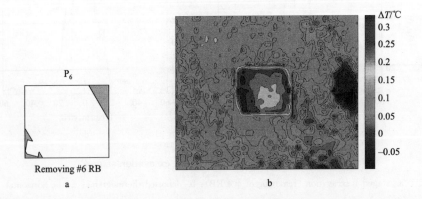

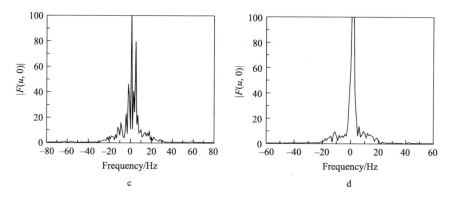

Figure 4.24　Figure set for the excavation stage P_6
a. staged 6 excavation (removing of #6 RB); b. denoised thermograms; c. the horizontal spectrum $|F(u, 0)|$; d. the vertical spectrum $|F(0, v)|$

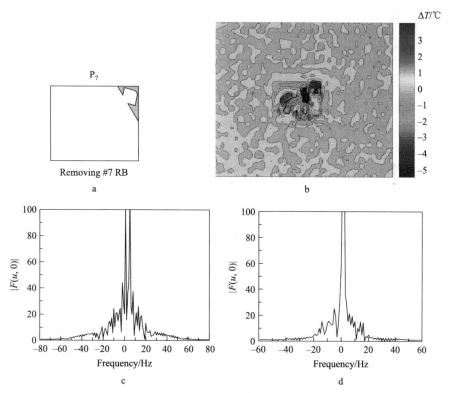

Figure 4.25　Figure set for the excavation stage P_7
a. staged 7 excavation (removing of #7 RB); b. denoised thermograms; c. the horizontal spectrum $|F(u, 0)|$; d. the vertical spectrum $|F(0, v)|$

It is seen from the thermal images from Figure 4.19b-Figure 4.25b that, during the staged excavation, the macroscopic displacement-induced frictional sliding involves: ① the elemental slabs; ② the rock layers and strata; ③ newly produced fractures of as mode II cracks when the local stress exceeding the elastic limit of the material.

4.7 Discussion

4.7.1 Excavation in differently inclined rocks over full-face excavation

Up to now, the published papers concerning the thermovision-based excavation experiments in the large-scale geological models in our institute include the excavation in horizontally inclined strata (He et al., 2010a; Gong et al., 2013), 45° inclined strata (He, 2011), 60° inclined strata (Gong et al., 2013) and vertically inclined strata (He et al., 2010b). Considering the length, two typical excavation statuses are compared here including the step 2 tunneling during the full-face excavation and the stage 3 excavation during the staged excavation. Figure 4.26-Figure4.29 show the figure sets for the excavation step 2 during the full-face excavation for the geomechanical model tests on excavation in 0°, 45°, 60° and 90° inclined stratified rock masses respectively.

Common image features of the thermal images obtained from the four excavations (see Figure 4.26b-Figure 4.29b) are the scattering-random IRT distribution exhibiting the thermoelasticity. Common spectral features (see Figures 4.26c、d and 4.29c、d) for the four excavations lie in the components with small amplitude distributed only in the low-or middle-wave bands, representing the fact that there no marked EDZ occurred at the step 2 excavation.

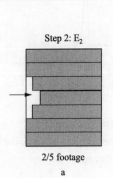

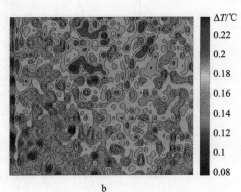

a   b

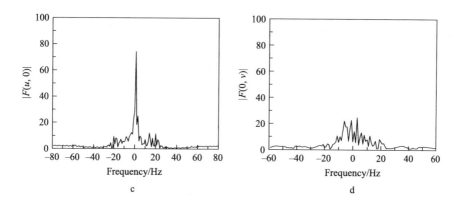

Figure 4.26 Figure set for the step 2 excavation
a. horizontally inclined strata; b. thermogram E_2 is from reference (Gong et al., 2013);
c. the horizontal spectrum $|F(u, 0)|$; d. the vertical spectrum $|F(0, v)|$

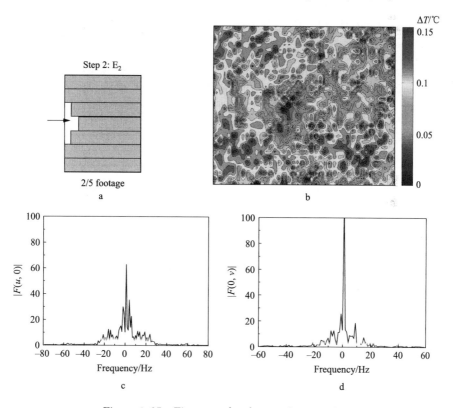

Figure 4.27 Figure set for the step 2 excavation
a. 45° inclined strata; b. thermogram E_2 is from reference (He, 2011); c. the horizontal
spectrum $|F(u, 0)|$; d. the vertical spectrum $|F(0, v)|$

Major differences of the thermal behavior for the four strata at step 2 tunneling during the full-face excavation depicted in the thermograms include:

(1) *Horizontal strata* (Figure 4.26): the scattered IRT in the mid-upper part of the thermogram representing the heterogeneous of the thermal response. That is, the horizontal strata behave like an isotropic continuum solid subject to the external loads and the heterogeneity comes from the stress concentration in such microscopic structures as natural flows and joints.

(2) *45° inclined strata* (Figure 4.27): weak anisotropy of the thermal behavior was indicated by the strata-paralleled IRT distribution, but the IRT strip is not persistent showing the static frictional effect are not strong.
60° inclined strata (Figure 4.28): strong anisotropy of the thermal behavior was illustrated by the persistent IRT strips.

(3) *Vertically inclined strata* (Figure 4.29): strong anisotropy of the thermal behavior was illustrated by the persistent IRT strips.

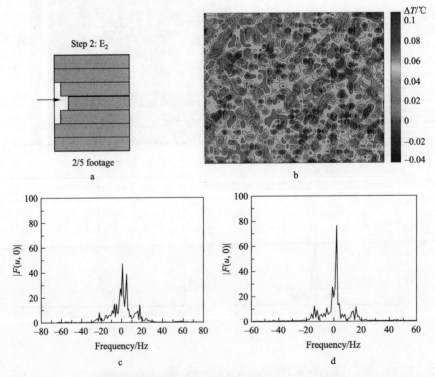

Figure 4.28 Figure set for the step 2 excavation
a. 60° inclined strata; b. thermogram E_2; c. the horizontal spectrum $|F(u, 0)|$;
d. the vertical spectrum $|F(0, v)|$

In simple word, at the state of thermoelasticity during the early excavation phase, the horizontal strata showed weak heterogeneous, the 45° inclined strata showed a weak anisotropy by the inconsistent static frictional bands, and the 60° inclined strata and vertical strata showed a strong anisotropy by the consistent static frictional bands.

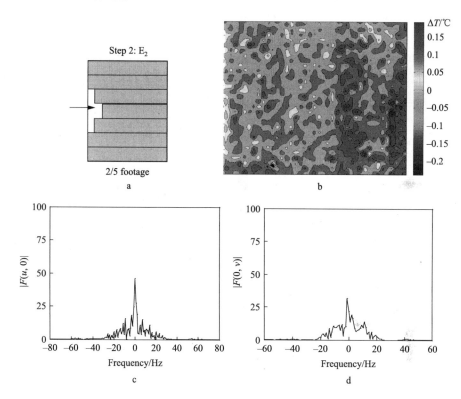

Figure 4.29 Figure set for the step 2 excavation
a. 90° inclined strata; b. thermogram E_2 is from reference (He et al., 2010b); c. the horizontal spectrum $|F(u, 0)|$; d. the vertical spectrum $|F(0,v)|$.

4.7.2 Excavation in differently inclined rocks over the staged excavation

Figure 4.30-Figure 4.32 show the figure sets for the excavation stage 3 during the staged excavation for the geomechanical model tests on excavation in 0°, 45°, 60°and 90° inclined stratified rock masses respectively.

Main features of the thermal behaviors for the four physical models can be compared as:

(1) *Horizontal strata* (Figure 4.30): localized-heterogeneous plastic damage around the face;

(2) 45° *inclined strata* (Figure 4.31): localized plastic damage with weak slip bands indicated by the localized IRT distribution and non-persistent IRT strips.

(3) 60° (Figure 4.32) *and vertically inclined strata* (Figure 4.33): strong slip bands constitute the major damage pattern represented by the persistent IRT strips parallel to the rock layers.

(4) The strongest wave propagation occurred in the 60° inclined strata, the second is in the 45° inclined strata, the third is in the vertical strata, and the horizontal is the weakest (He et al., 2010b). The intensity of the stress wave propagation corresponds to the severity of the excavation-induced damage.

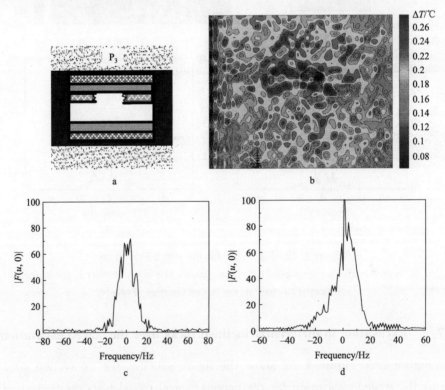

Figure 4.30 Figure set for stage 3 excavation in horizontal strata
a. horizontally inclined strata; b. thermogram P_3 is from reference (Gong et al., 2013);
c. the horizontal spectrum $|F(u, 0)|$; d. the vertical spectrum $|F(0, v)|$

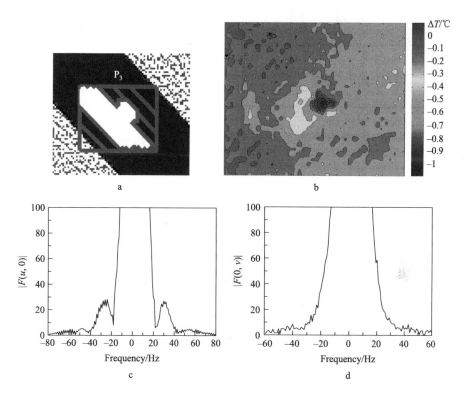

Figure 4.31 Figure for stage 3 excavation in 45° inclined strata
a. 45° inclined strata; b. thermogram P_3 is from reference (He, 2011); c. the horizontal spectrum $|F(u, 0)|$; d. the vertical spectrum $|F(0, v)|$

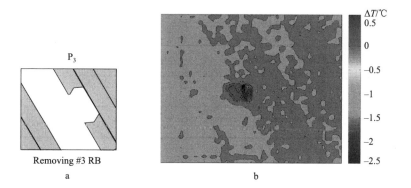

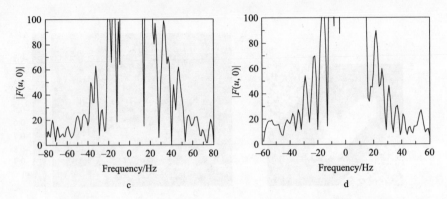

Figure 4.32 Figure set for stage 3 excavation in 60° inclined strata
a. 60° inclined strata; b. thermogram P_3; c. the horizontal spectrum $|F(u, 0)|$;
d. the vertical spectrum $|F(0, v)|$

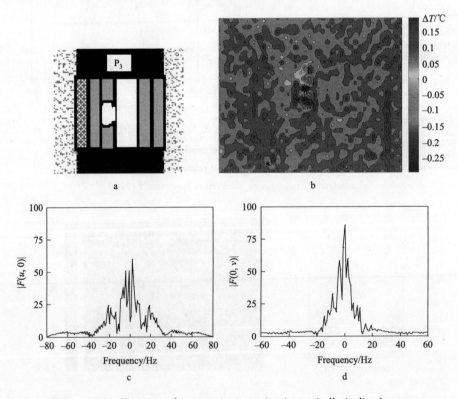

Figure 4.33 Figure set for stage 3 excavation in vertically inclined strata
a. 90° inclined strata; b. thermogram P_3 is from reference (He et al., 2010b); c. the horizontal
spectrum $|F(u, 0)|$; d. the vertical spectrum $|F(0, v)|$

4.7.3 Summary

The procedure for processing of a thermogram was presented involving removal of the environmental radiation noise by image subtraction, suppression of the salt-and-pepper noise by using a *median* filter, and eliminating of periodical noise by using a Gaussian-high-pass filter.

Borders and edges of the excavation-induced frictional sliding bands with changed stress represented by the changed brightness can be distinguished clearly in the processed thermograms by utilizing the developed algorithm. The denoised thermograms are helpful to better understand different thermal behaviors of the rock under excavation and the proposed algorithm may be applicable in processing other noise-contained thermograms.

Such image features as mean of the temperature field $\langle IRT \rangle$, horizontal and vertical spatial Fourier spectra, and denoised thermograms were extracted. Based on analysis of these features, different thermal-mechanical coupling processes as the reversible thermo-elastic coupling and intrinsic dissipation for the 60° inclined strata under excavation are discriminated, that is:

(1) Pure elastic represented by scattering-random IRT distribution with no major wave components such as E_0 and E_1.

(2) Elastic with weak static friction indicated by scattered IRT strip with dim contrast with minor wave components such as E_2.

(3) Elastic with strong static friction depicted by higher contrast of the IRT strips with more wave components with major wave components such as E_3-E_6.

(4) Plastic with static friction at critical state represented by high $\langle IRT \rangle$ with a sharp drop, localized IRT distribution and dim slip bands with more major wave components such as P_2.

(5) Dynamic frictional sliding indicated by the highest $\langle IRT \rangle$ with a significant drop, brightness of the slip bands, with most number of the major wave components such as P_3.

(6) Localized dynamic frictional sliding depicted by the low $\langle IRT \rangle$, and brightness of the slip bands, with less number of the major wave components such as P_4.

(7) These thermal precursors can be used for the assessment of the excavation damage and prediction of the deterioration of the rock strength.

Comparison of this research on the thermal responses to the tunnel excavation with the published excavation experimental studies in our institute were made and the major results, for example, can be concluded as: ① for step 2 tunneling during the full-face excavation, the horizontal strata (He et al., 2010a) behaves like continuum with some weak heterogeneity; the 45° inclined strata (He, 2011) shows weak anisotropy by the inconsistent IRT strips; the 60° and vertical strata (He et al., 2010b) exhibit a strong anisotropy by the persistent IRT strips. ② for stage 3 during the staged excavation, the horizontal strata has a localized excavation around the face; the 45° inclined strata has a weak slip bands with localized damage around the face; and the 60° inclined and vertical strata have strong slip bands as the major damage form.

References

Barla M. 2008. Numerical simulation of the swelling behavior around tunnels based on special triaxial tests. Tunnelling and Underground Space Technology, 23: 508-521.

Brady B G H, Brown E T. 2004. Rock Mechanics for Underground Mining. New York: Kluwer Academic Publishers.

Cai M, Kaiser P K. 2005. Assessment of excavation damaged zone using a micromechanics model. Tunnelling and Underground Space Technology, 20: 301-310.

Carpinteri A, Paggi M. 2005. Size-scale effects on the friction coefficient. Int. J. Solids Struct, 42: 2901-2910.

Cheon D S, Jeon S, Park C, et al., 2011. Characterization of brittle failure using physical model experiments under polyaxial stress conditions. International Journal of Rock Mechanics and Mining Science, 48: 152-160.

Connolly M, Copley D. 1990. Thermographic inspection of composite material. Mater. Evaluation, 8(12): 1461-1463.

Dai, H Y, Guo J, Yi S H, et al. 2013. The mechanism of strata and surface movements induced by extra-thick steeply inclined coal seam applied horizontal slice mining. J. Chin. Coal Society, 38(7), 1109-1115.

Gong W L, Gong Y X, Long Y F. 2013. Thermography analysis of a roadway excavation experiment in 60° inclined stratified rocks. International Journal of Rock Mechanics and Mining Science, 60: 134-147.

Gong W L, Zhao H Y, An L Q, et al. 2008. Temporal and spatial analysis of infrared images from water jet in frequency domain based on DFT. J. Beijing Univ. Aeronaut. Astronaut, 34(6): 690-694.

Gong Y X, Long A F, Gong W L, et al. 2012. Infrared thermal imaging and image processing of turbulent jet. Infrared, 33(5): 42-47.

Gonzalez R C, Woods R E, Eddins S L. 2005. Digital Image Processing. Beijing: Publishing

House Electronics Industry.

Grinzato E, Marinetti S, Bison P G, et al. 2004. Comparison of ultrasonic velocity and IR thermography for the characterization of stones. Infrared Phys. Technol, 46; 63-68.

He M C, Gong W L, Zhai H M, et al. 2010a. Physical modeling of deep ground excavation in geologically horizontally strata based on infrared thermography. Tunnelling and Underground Space Technology, 25: 366-376.

He M. C. 2011. Physical modeling of an underground roadway excavation in geologically 45° inclined rock using infrared thermography. Eng. Geol, 121: 165-176.

He M C, Jia X N, Gong W L, et al. 2010b. Physical modeling of an underground roadway excavation vertically stratified rock using infrared thermography. International Journal of Rock Mechanics and Mining Sciences, 47: 1212-1221.

Jeon S., Kim J., Seo Y, et al., 2004. Effect of a fault and weak plane on the stability of a tunnel in rock—a scaled model test and numerical analysis. International Journal of Rock Mechanics and Mining Sciences, 41(1): 658-663.

Jia P, Tang C A. 2008. Numerical study on failure mechanism of tunnel in jointed rock mass. Tunnelling and Underground Space Technology, 23: 500-507.

Ju W J, Li Q, Wei D, et al. 2006. Pressure character in caving steep-inclined and extreme thick coal seam with horizontally grouped top-coal drawing mining method. J. Chin. Coal Society, 31(5), 558-561.

Kamata G, Mashimo H. 2003. Centrifuge model test of tunnel face reinforcement by bolting. Tunnelling and Underground Space Technology, 18(2): 205-212.

Luong M P. 1990. Infrared thermovision of damage processes in concrete and rock, Eng. Fracture Mech, 35(1/2/3): 291-310.

Luong M P. 1995. Infrared thermographic scanning of fatigue in metals. Nuclear Eng. Design. 158: 363-376.

Luong M P. 2007. Introducing infrared thermography in soil dynamics. Infrared Phys. Technol. 49: 306-311.

Meguid M A, Saada O, Nunes M A, et al. 2008. Physical modeling of tunnels in soft ground: review. Tunnelling and Underground Space Technology, 23: 185-198.

Mukhopadhyay S, Chanda B. 2000. A multiscale morphological approach to local contrast enhancement. Signal Process, 80: 685-696.

Pastor M L, Balandraud X, Grédiac M, et al. 2008. Applying infrared thermography to study the heating of 2024-T3 aluminum specimens under fatigue loading. Infrared Phys. Technol, 51: 505-515.

Pratt W K. 1991. Digital Image Processing. New York: Wiley.

Read R S. 2004. 20 years of excavation response studies at AECL's Underground Research Laboratory. International Journal of Rock Mechanics and Mining Sciences, 41: 1251-1275.

Sharma J S, Bolton M D, Boyle R E. 2001. A new technique for simulation of tunnel excavation in a centrifuge. Geotech. Test. J, 24(4): 343-349.

Sitharam T G, Latha G M. 2002. Simulation of excavations in jointed rock masses using a practical equivalent continuum approach. International Journal of Rock Mechanics and Mining Sciences, 39: 517-525.

Steinberger R, Valadas-Leitão T I, Ladstätter E, et al. 2006. Infrared thermographic techniques for non-destructive damage characterization of carbon fiber reinforced polymers during tensile fatigue testing. Int. J. Fatigue, 28: 1340-1347.

Tanaka T, Sakai T. 1993. Progressive failure and scale effect of trapdoor problem with granular materials. Soils Found, 33(1): 11-22.

Tang Y G, Kung G T C. 2009. Application of nonlinear optimization technique to back analyses of deep excavation. Compt. Geotech, 36: 276-290.

Varadarajan A, Sharama K G, Desai C S, et al. 2001. Analysis of a powerhouse carven in the Himalayas. Int. J. Geomech, 1(1): 109-27.

Wu L X, Liu S J, Wu Y H, et al. 2002. Changes in infrared radiation with rock deformation. International Journal of Rock Mechanics and Mining Sciences, 39; 825-831.

Wu L X, Liu S J, Wu Y H, et al. 2006a. Precursors for rock fracturing and failure-Part II: IRRT-Curve abnormalities. International Journal of Rock Mechanics and Mining Sciences, 43: 483-493.

Wu L X, Liu S J, Wu Y H, et al. 2006b. Precursors for rock fracturing and failure-Part I: IRR image abnormalities. International Journal of Rock Mechanics and Mining Sciences, 43: 473-482.

Wu L X, Wu Y H, Liu S J, et al. 2004. Technical Note: Infrared radiation of rock impacted at low velocity. International Journal of Rock Mechanics and Mining Sciences, 41; 321-327.

Young R P, Collins D S. 2001. Seismic studies of rock fracture at the underground Research Laboratory, Canada. International Journal of Rock Mechanics and Mining Sciences, 38(6): 787-799.

Young R P, Collins D S, Reyes-Montes J M, et al. 2004. Quantification and interpretation of seismicity. International Journal of Rock Mechanics and Mining Sciences, 41: 1317-1327.

Zimmerman R W, Al-Yaarubi A, Pain C C, et al. 2004. Non-linear regimes of fluid flow in rock fractures. International Journal of Rock Mechanics and Mining Sciences, 41(3): 384-384.

Chapter 5　Excavation in 45° strata

5.1　Introduction

Understanding the mechanism of rock damage in the creation and operation of an underground cavern in jointed rock masses has been a topic of research for engineers in various fields, including developing transportation tunnels, petroleum production drillings, and underground mines, for example (He, 2006).

Extensive researches with varied methods, including numerical, laboratory experiments, and in-situ tests, have been conducted to investigate the excavation related problems. In-situ tests involve the investigation of the excavation responses (Read, 2004); monitoring the displacement of a powerhouse cavern during excavation (Li et al., 2008), back analysis of excavation-induced wall deflection (Tang and Kung, 2009), etc.

In recent decades, numerical methods have increasingly been used in simulating the stress redistribution field created by the tunneling impacts (Cai, 2008), the swelling behavior around tunnels (Barla, 2008); acoustic emission in underground excavations (Cai et al., 2007); failure mechanism of tunnel in jointed rock mass (Jia and Tang, 2008).

Although computational techniques have progressed fast, geotechnical researchers depend heavily on physical modeling tests with a reduced model have been widely used at laboratory in investigating the failure mechanism and behavior of tunnel face (Chambon and Corte, 1994); the stability of the face and unsupported span over tunnel excavation in weak rock (Lee and Schubert, 2008); ground movement and collapse mechanisms induced by tunneling in clayey soil (Wu and Lee, 2003); progressive failure and scale effect of trapdoor problem (Tanaka and Sakai, 1993); tunnel face reinforcement by bolting with centrifuge model test (Sharma et al., 2001; Kamata and Masimo, 2003).

Field experiments and in-situ tests are limited both in time and costs. Numerical modeling allows one to conduct more realistic analyses that take into account the complexity of the jointed rock masses except for rock masses containing natural discontinuities of varying size, strength and orientation. In practice, it is almost impossible to explore all of the joint systems or to investigate all their mechanical characteristics and implementing them explicitly in a theoretical model (Sitharam and Latha, 2002).

The physical modeling tests at laboratory with a small-scale model are also limited by the fact that in situ stresses are not realistically simulated, and there is inconsistency of scaling factors for different measured quantities, e. g. length, inertia force, creep (Meguid et al., 2008). Due to the increase in mining depth and in urbanization found all over the world, so many tunnels were and will be built under complex geological conditions. As a result, a comprehensive and deepened understanding of the tunneling induced displacements and stresses and their impact on nearby structures have been becoming imperative.

To achieve these goals, new physical modeling methods for simulating the excavation related problems in a large-scale geological model, incorporated with the state-of-the art measuring techniques that is capable of visualization of the stress field in real-time and over the entire field with a nondestructive manner around the tunneling face, should be introduced into the laboratory model test.

The objective of this experiment presented in Chapter 5 is to understand the dynamical rock responses to the tunnel excavation in the geologically 45° inclined rock strata. Infrared thermography was employed for conducting the geomechanical test and capturing the thermal responses of the surrounding rocks under excavation. The resultant infrared thermography series were then utilized in observing and capturing the mechanical and structural behaviors of the rocks under the roadway tunnel excavation.

5.2 Short review of infrared detection

Although some contact detection technologies, such as acceleration sensors, strain gauges, stress wave propagation meters and displacement measurement, have been successfully applied both in situ and laboratory, but it is not suited for such dynamical processes as rock impaction because of the intrusive disturbance and damage-prone problem of the contacting detectors (Shi et al., 2007).

The widely used non-destructive detection technique in rock mechanics includes acoustic emission (AE) and electromagnetic radiation. It describes the progressive development failure of the rock indirectly by the transformation or statistics of the test data sets (Majewska and Mortimer, 1998; Shiotani, 2006; He et al., 2010c).

Owing to the thermomechanical coupling, infrared thermography provides a non-destructive, non contact and real-time test to observe the physical process of material degradation and to detect the occurrence of intrinsic dissipation without surface contact or in any way influencing the actual surface temperature of the tested object (Luong, 1995). It produces heat images directly from the invisible radiant energy (dissipated energy) emitted from stationary or moving objects at any distance. By using an infrared thermographic detection system, one can observe and capture the physical and structural changes of an object, represented by the infrared radiation temperature changes, in real time and over the entire field of the viewed surface.

Abnormal rising of the surface radiation temperature were observed on the infrared thermal images in the climate satellites several days before and after earthquakes. It was recognized that the earthquakes is related to the changes of the stress field of the earth's crust, and then the change of the stress field causes electromagnetic radiation including the infrared radiation which has the thermal effect (Qiang et al., 1990).

The fact that infrared radiation energy varies with the change of stress field of the loaded rocks was further verified by the subsequent studies: uniaxial and biaxial loading of sedimentary rocks and igneous rocks (Cui et al., 1993; Zhi et al., 1996); the structural stress fields of rock and gas burst disasters resulting in local abnormal high temperature in mine (Wu and Wang, 1998); projectile impact on rock (Shi et al., 2007); the roadway tunnel under plane loading (He et al., 2009) and its excavation process (He et al., 2010a, 2010b) in the physical analogue models in a horizontal and vertical strata consisting of alternating layers of sandstone, mudstone and coal.

The thermal effect caused temperature increment ΔT is affected by the combined factors, i.e., thermal-elastic effect, rock fracturing, rock friction, environmental radiation and heat transfer. As the physical model in our test is large in size, the heat transfer between the loading platen and the analogous model is small and the same for the environmental radiation, the loading process can be considered adiabatic, the heat transfer and the environmental radiation can be neglected.

The infrared temperature increment $\Delta T(K)$ can, therefore, be described qualitatively by the following equation (Wu and Wang, 1998):

$$\Delta T = \Delta T_1 + \Delta T_2 + \Delta T_3 \tag{5.1}$$

The terms at the right-hand side of Eq. (5.2) can be explained in what follows:

(1) ΔT_1 is infrared temperature increment due to the thermal-elastic effect, K, positive or negative, which has a relation for a plane loading state (Wu et al., 2006a, 2006b):

$$\Delta T_1 = \gamma \beta^{-1} T \Delta (\sigma_1 + \sigma_2) \tag{5.2}$$

where, T is the physical temperature of rock, K; β is a constant correction factor related to rock surface radiant emittance, rock thermoelastic fracture, MPa · K/V; γ is a transfer factor between the detected voltage signal and the infrared temperature, K/V. According to thermoelastic theory, the change of surface temperature of any solid unit is linearly related with the change in the sum of its three principal stresses $\Delta(\sigma_1 + \sigma_2)$, MPa (Harwood and Cummings, 1991). If $\Delta(\sigma_1 + \sigma_2) > 0$, ΔT_1 will be positive and vice versa.

(2) ΔT_2 is the temperature decrement due to the expansion of the initial fissures, joints and newly produced fractures, K, negative. The production of new fracture needs to consume energy, and the extension of fissures, joints and fractures also needs to consume energy. Hence, ΔT_2 is always negative. The more the production and extension, the more the negative effect of ΔT_2.

(3) ΔT_3, is the temperature increment due to the friction heat, K, positive. The friction heat is produced by the friction action between joints, fissures and grains of deforming rock when subject to the external load. The heat induced by the friction is always positive, and monotonically increases with the increasing deformation and newly produced fractures.

Identical or similar results on thermal effect were achieved by such infrared thermography based studies as: thermographic inspection of loaded composite material (Connolly and Copley, 1990; Steinberger et al., 2006); loaded metals (Luong, 1995; Pastor et al., 2008); and loaded rocks and rock-like materials (Brady and Rowell, 1986; Luong, 1990; Qiang et al., 1990; Cui et al., 1993; Zhi et al., 1996; Geng et al., 1998; Wu and Wang, 1998; Wu et al., 2002, 2004, 2006a, 2006b; Grinzato et al., 2004; Shi et al., 2007). These well-established thermomechanical coupling knowledge were utilized as references in the analyzing and interpreting the heat images in this study.

5.3 Experiment

5.3.1 Model construction

As discussed in Chapter 3, the phototype simulated in the geomechanical model test is the excavation of a roadway tunnel in **QI SHAN** underground coal mine in Xuzhou coal mining district, located in Jiangsu province, eastern China. The roadway tunnels in this underground coal mine are at depths ranging from 300 m to more than 1000 m below the ground surface (note that the "depth" in the following context is referred to counting from the ground surface). The main exposure lithologies are sandy mud rock, mudstone and sandy rock. In this chapter, the physical modeling of roadway excavation in geologically 45° inclined stratified rocks will be introduced.

Major rock types of the sedimentary rocks for the prototype are sandstone, mudstone and coal seam. Unconfined uniaxial compressive tests and basis laboratory experiments were conducted on the rock specimens sampled from the surrounding rock masses of the field case. The obtained results are reported in Table 3.1, which are the real rock properties simulated by the artificial materials in our test. From Table 3.1 one can see that the unconfined uniaxial compressive strength (UCS) of the coal rock is close to the threshold strength 25 MPa for the soft rocks at depth (He et al., 2002).

By utilizing the similarity principles, based on the dimensional analysis, the modeling materials were produced as reported in Table 3.2. Geomechanical model consisting of alternating rock layers of sandstone, mudstone and coal seam were constructed using the PFESA method and geological model shown in Figure 3.8. The dimension of the physical model is 1600 mm × 1600 mm × 400 mm, and the roadway excavation zone was designed as a cubic space with a face area of 250 mm × 200 mm and tunneling length of 400 mm. Figure 5.1a and b show schematically the model size and simulated engineering rock mass by the geometrical scale factor $a_l = 12$ respectively.

The geomechanical model was constructed with total nine strata, including one sandstone stratum, four mudstone strata and four coal seam strata as shown

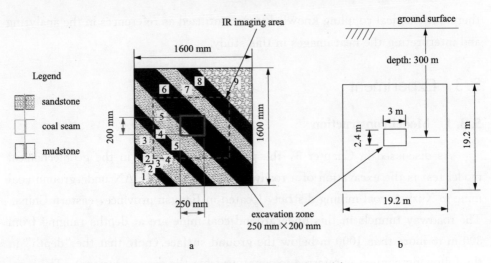

Figure 5.1 The tested PFESA model
a. schematic of the model; b. rock reality of the underground cavern simulated by the PFESA model

in Figure 5.1. All the strata were inclined at an angle of 45° with respect to the horizontal, to simulate the geologically 45° inclined real rock strata. These model rock strata, indexed 1-9 from the left to the right in the physical model, were assembled in accordance with the sequence of the geological section of the QI SHAN coal mine. Rock properties, geometrical parameters and used elementary slabs are reported in Table 5.1. The excavation zone is located in the stratum 6, the thickest coal seam among the strata (see Figure 5.1a).

Table 5.1 Geological section of the geomechanical model

Geological section	Stratum No.	Rock types	Layer thickness/mm	Layer numbers
	1	sandstone	440	14
	2	coal seam	140	10
	3	mudstone	120	7
	4	coal seam	250	17
	5	mudstone	150	5
	6	coal seam	60	4
	7	mudstone	140	5
	8	coal seam	60	3
	9	mudstone	240	8

5.3.2 Testing procedure

1. Excavation scheme

The roadway excavation zone was located on the stratum 6 (coal seam) centered on the PFESA model. The tunnel excavation space in the model was a cubic volume measuring 250 mm × 200 mm × 400 mm. The total roadway excavation volume was divided into seven sub-spaces along the strike of the strata as seen in Figure 5.1a. These sub-spaces being excavated were referred to as "rock block (RB)" as shown in Figure 5.2. Figure 5.2a is the front face of the excavation zone and 5.2b is the 45° rotary sectional view of the roadway tunnel. The RBs were numbered #1-#7 corresponding to the roadway excavation sequence.

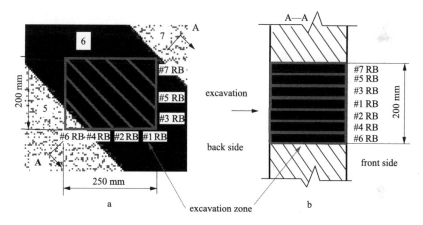

Figure 5.2 Roadway excavation scheme
a. front face of the excavation zone on the model; b. 45° rotary sectional view of the excavation zone

Same as the excavation plan in horizontal strata introduced in chapter 4, two phases of excavation were planned, i.e, the full-face excavation, i.e., tunneling on the #1 RB and the adjacent rocks until a small passage is cut through; and phase 2: staged excavation, i.e., removing one RB at each tunneling stage. For the full-face excavation, the excavated volume of the artificial materials from the roadway tunnel was referred to as "footage" thereafter. For the staged excavation, the term "excavation stage" denotes the removing each of the RBs, and total of seven excavation stages were performed step by step in sequence during the phase 2 excavation.

2. Excavation method

The excavation of the roadway tunnel without support was started at the back side of the model, and went through to the front face, with a chisel and a hammer as the tunneling tools, simulating the tunnel excavation by drill and blast method. Detailed description of the excavation method can be found in chapter 4 (see Figure 4.6).

3. Boundary condition

During the excavation, the vertical load and lateral load were applied on the model (simulated as quasi-2D plane strain state) by using the testing frame (see chapter 4). As schematically shown in Figure 5.3, the testing frame was equipped with a hydraulic servo system, imposing the uniformly distributed load on the top and two side boundaries of the model. The bottom of the model was fixed on the basement of the machine, simulated as a rigid boundary condition during the excavation.

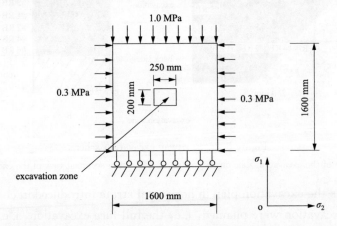

Figure 5.3 Simulated boundary stress condition over the roadway excavation

During the excavation, the vertical stress σ_1 was kept at a constant value of 1.0 MPa, corresponding to the mining depth 300 m. We chose the lateral pressure coefficient $\lambda=0.3$, (defined by $\lambda=\sigma_2/\sigma_1$, σ_2 is the lateral stress), for reproducing the unbalanced stress state on boundaries in deep ground, and σ_2 was also kept constant at 0.3 MPa during the whole excavation processes as illustrated in Figure 5.3. It is noted that the boundary condition is the same as that in the excavation in 45° inclined stratified rocks (see chapter 4).

5.4 Infrared detection

5.4.1 Infrared thermography

Infrared thermography of TVS-8100 MK II model was used in our test for detection of the infrared radiation field on the excavation zone in the physical model. The specifications for the thermography were reported already in chapter 4 and chapter 5. The experimental setup is shown in Figure 5.4. Figure 5.4a is the photograph of the experiment setup and Figure 5.4b is the schematic. For 24 hours earlier prior to the testing, all the instruments were placed in the same room with the PFESA model, so that the detected infrared temperatures are the temperature variation due to the excavation impact. The 400 mm×400 mm mark frame of red colored plastic thin membrane was attached with glue to the model front face centrosymmetric to the roadway cross section, indicating infrared imaging area, as seen in Figure 5.4b. The infrared camera and video camera were placed at the model's front face while the excavating operations started from the rear face.

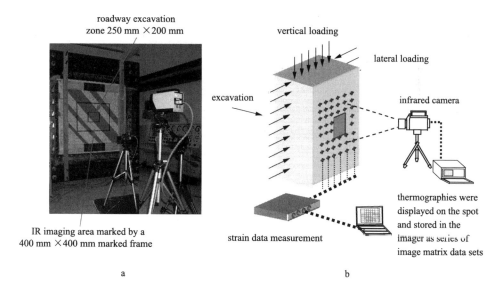

Figure 5.4 Experimental test setup
a. photograph of the infrared camera and the excavation area in view; b. the diagram for the testing system

5.4.2 Energy release index

Since thermal image is the visualization of the infrared radiation temperature (IRT) distribution on the surface in view, its pixel matrix $f(i, j)$, where i and j are the number of pixels; $i=1,2,\cdots, M$; $j=1,2,\cdots, N$, is IRT data set at the specific instant. From a stochastical point of view, the image matrix $f(i, j)$ can be regarded as a two-dimensional stochastic field.

The sample mean of the image matrix can be calculated by:

$$\langle f_{ij} \rangle = \frac{1}{M} \frac{1}{N} \sum_{i=1}^{N} \sum_{j=1}^{M} f(i,j) \tag{5.3}$$

where, $\langle \cdot \rangle$ represent the sample mean; $M \times N = 120 \times 160$, is the total number of the pixels in one frame of the infrared image series. The mean value $\langle f_{ij} \rangle$ represents the transient infrared power intensity emitted from the target at time of sampling. Thus $\langle f_{ij} \rangle$ is termed *"energy release index"* and is expressed by $\langle IRT \rangle$ as a result of its real meaning.

Time series $\langle f_{ij} \rangle_k$ can be obtained by computing the sample mean using Eq. (5.3) on the thermal sequence $f(i, j)_k$ ($k=0,1,2,\cdots$ is the number of infrared images), and k can be converted into the discrete time variable t by multiplying the variable k with the imaging frequency used in the test.

The first frame of the thermal image was taken when the model was at the initial state, and used as the basic reference for the calibration of the temperature changes of the following infrared images. Let

$$\bar{T}_k = \langle f_{ij} \rangle_k, \ k = 0,1,2,\cdots$$

normalized $\bar{T}_k$ is defined by

$$T_k = \frac{T_0 - \bar{T}_k}{T_0 - T_{max}} \tag{5.4}$$

In the following context in this paper, T_k is actually the normalized $\langle IRT \rangle$ and is referred to as *"normalized energy release index"* or *"normalized IRT or IRT"* in the following context, for sake of simplicity. The *normalized* IRT represents the IRT variations relative to the initial state of the model or the energy dissipation level relative to the maximum IRT mean value T_{max}. Therefore, IRT, was used as a measure of the rock responses to the excavation processes.

5.4.3 Image processing algorithm

For removing the noises contained in the raw thermal images as discussed in chapter 4, following algorithm are proposed and listed in the processing sequence:

(1) subtraction of the first frame from the follow-up infrared images for eliminating the background radiation noises;

(2) performing the median filter for suppressing the impulsive noise;

(3) performing the Gaussian-high-pass filter, i. e. , GHPF, for removing the additive-periodical noise.

The above image processing operations were realized in the MATLAB 8.0 platform based on the Image Processing Toolbox (IPT) functions in the Matlab macro code.

5.4.4 Principles for image analysis

Infrared image represents the precursory information about rock fracture by pseudo-colors, borders and edges and light or dark regions which are the different temperature distribution zones.

For facilitating comprehension of the infrared images, following well-established rules can be used:

(1) hot or positive colors stand for high-level temperatures and cool or negative colors for low-level temperatures;

(2) high temperature indicates high stress level due to friction, shear or stress concentration, and low temperature indicates low stress level due to tensile cracking, stress release or unloading;

(3) temperature distribution scale in the light or dark regions may indicate the scale of the rock damage and localized temperature distribution represents plastic deformation or permanent damage;

(4) the borders and edges that separate hot-and cool-colored zones illustrate the modes of rock behavior.

5.4.5 Fourier analysis

When rock masses undergone excavation impact, some of the strain energy in the system is converted into kinetic energy that needs to be dissipated. As the excavation advances, the dynamic loading and unloading conditions exist, which will be generated the stress waves. The excavation induced stress wave will travel in the rock layers and reflect at the interfaces of the rock boundaries.

Stress wave propagation can be described in the frequency domain by Fourier transformation of the infrared images. We resample the two 1-dimensional (1-D) data sets, i.e., x_i, $i=1,2,\cdots,120$ and y_j, $j=1,2,\cdots,160$, along the vertical and horizontal axes right across that center of the infrared image and performed one-dimensional (1-D) Fourier transformation upon them, which were referred to as the "horizontal spectrum" and "vertical spectrum" respectively. Figure 5.5 illustrates the diagram of the sampling and performing Fourier transformation from the thermography collected at the instant E_1 in the phase 1 excavation.

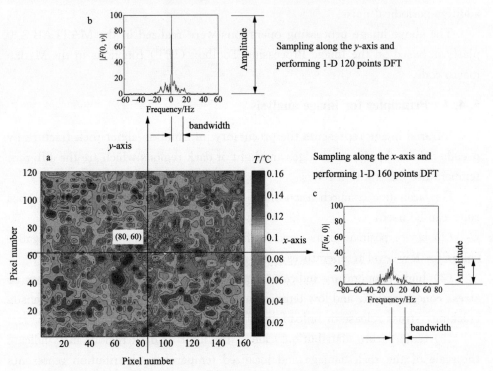

Figure 5.5 Illustration of the Fourier analysis of the infrared image: the image matrix was resampled along the horizontal and vertical axes passing through the center of the image
a. the vertical spectrum was obtained by 120 points DFT; b. the vertical sepctrum obtained by 120 points DFT; c. the horizontal spectrum was obtained by 160 points DFT; thermography (a) is with respect to E_1

The 1-D Fourier transformation was achieved by the DFT (Discrete Fourier Transform) algorithm (Pinsky, 2003):

$$F(u) = \text{DFT}[x_i] = \sum_{i=0}^{M} x_i \exp[-2\pi i(ui/M)] \tag{5.5}$$

and

$$F(v) = \text{DFT}[y_j] = \sum_{j=0}^{N} x_i \exp[-2\pi j(uj/N)] \tag{5.6}$$

where, u is the discrete horizontal frequency variable, Hz; v is the discrete vertical variable ($u = 1, 2, \cdots, M$; $v = 1, 2, \cdots, N$), Hz; $F(u)$ and $F(v)$ are the horizontal and vertical spectra respectively.

In our analysis of the thermographers, the amplitudes of the horizontal and vertical spectra, i.e., $|F(u)|$ and $|F(v)|$, were used to characterize the directional propagation of the excavating induced stress waves along the horizontal and vertical respectively. In the spectral-frequency analysis, we asses propagating stress waves according to the level of the amplitude, and the "bandwidth" or "band" in which the high level amplitude distributes. The detailed introduction of performing a spectral-frequency analysis on the infrared thermography can be found in the literatures (He et al., 2010a).

5.5 Results and Discussions

5.5.1 Overall thermal response

IRT, the intensity of the infrared radiation in essence, was used as a measure in characterizing the rock behaviour during the excavation processes. Figure 5.6 shows the time-marching scheme of IRT over the two roadway excavation processes, and Figure 5.7 illustrates the tunneling face advancement of the excavation steps over the full-face (phase 1) excavation, and Figure 5.8 depicts the total of seven stages during the staged (phase 2) excavation. From Figure 5.6 we can see that the IRT curve has two different evolution patterns, i.e., piecewise linear increase with jump discontinuity points (referred to as "multi-linearity" thereafter) over the full-face excavation and fluctuation during the staged excavation.

We use E_0-E_6 and P_0-P_8, referred to as points of interests (POI) standing for the excavation steps and stages in the two excavation phases respectively, marked them on the IRT profile, as seem in Figure 5.6. POIs of E_0-E_6 corresponds to the full-face excavation (see Figure 5.7), E_0 is the intact state of the physical, E_1-E_5 standing for the excavation advancement from step1 to step 5, and E_6 representing the instant when a small passage was about to cut through (destruction of the #1 RB for the first time).

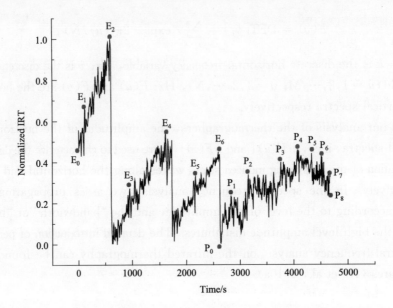

Figure 5.6   Profile of the measured IRT against the excavation advancement

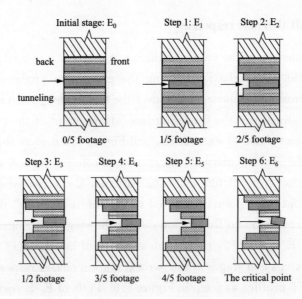

Figure 5.7   Diagram for the full-face (phase 1) excavation process at a sectional view

During the full-face excavation, the accumulation of the elastic energy caused the linear increase in IRT and the excavation induced rock sliding induced the jump discontinuity points like POI E_2 and E_4. The evolution pattern of IRT in this test is different from that in the physical modeling of the roadway tunnel excavation in the horizontal strata (He et al., 2010a), where the IRT has no jump discontinuity point during the full-face excavation for the horizontal strata. The IRT time-marching scheme depicted in Figure 5.6 validates the fact that for the jointed rock masses with steep dip angles, the excavation impact, although the tunnel face advancing shallow, will induce faulting damage between the rock layers.

It was noted that the IRT profile attains its global maximal point at POI E_2 during the very early excavations, and reaches its local maximal points at POIs E_4 and E_6 (see Figure 5.7). It indicates that the steep-angled rock layers are prone to slide along their faulted bedding planes and hard to store the elastic energy inside caused by the excavation impact. The time-marching scheme of the IRT profile is divided by POI E_6, i.e., before E_6 the IRT has a multi-linearity evolution pattern and after E_6 the IRT was change to fluctuate. As a result, POI E_6 is the "critical point" dividing the time-marching scheme of the IRT profile in the two excavation processes.

Immediately after POI E_6 (the critical point), a sharp drop was seen on the IRT curve at POI P_0, indicating the faulting-damage by cutting through a small passage in the #1 RB. During the staged excavation (see Figure 5.8), removal of one RB at each of the excavation stage caused a quasi-periodic loading and unloading impact on the strata resulting in the consequential fluctuation of the IRT profile. According to the time-marching scheme of the IRT curve, the rock behaviour during the full-face excavation was in a multi-linearity manner, and during the staged excavation in a plastic manner, from an energy dissipation point of view within the infrared radiation regime.

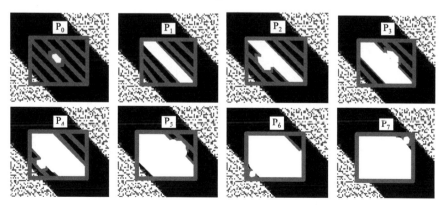

Figure 5.8 Diagram for the staged (phase 2) excavation process at a front view

5.5.2 Characterization of the full-face excavation

The enhanced understanding of excavation responses was achieved by characterization of the rock behaviour spatially using the denoised thermography and the directional stress wave propagation by the spectral-frequency analysis of the thermography. Figure 5.9-Figure 5.15 show the figure sets for E_0-E_6 during the full-face excavation respectively. In each figure set, the first figure is the excavation diagram (marked by letter a); the second is the thermal image indicated by b; the third and fourth are the horizontal spectrum $F|(u, 0)|$ (marked by c) and vertical spectrum $F|(0, v)|$ (marked by d), respectively.

Rock responses over the full-face excavation can be understood by characterization of the image features depicted in Figure 5.9-Figure 5.15. In Figure 5.9, the excavation diagram E_0 (Figure 5.9a) shows the intact state of the model. E_1 (Figure 5.10a) and E_2 (Figure 5.11a) show the face advancement at 1/5 and 2/5

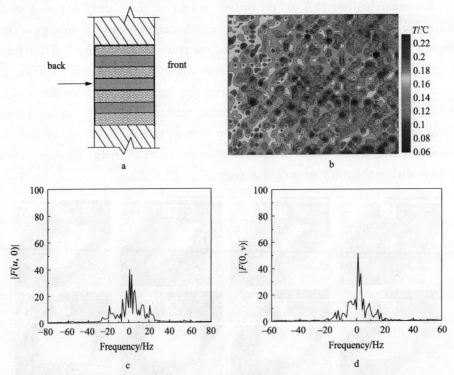

Figure 5.9 Figure set for the initial state E_0
a. the initial state; b. denoised thermograms; c. the horizontal spectrum $|F(u, 0)|$;
d. the vertical spectrum $|F(0, v)|$

footages, there was no sliding between the rock layers as the face went forward shallow away from the back surface of the model. E_3-E_5 (Figure 5.12a-Figure 5.14a) illustrate the face advancing deep into the model and rock layer sliding occurred. E_6 depicts the instant when the #1 RB was destructed (the critical point, corresponding to the global maximal value in the IRT curve). In the thermal image E_0 (Figure 5.9b), the random-scattering IRT distribution indicates the intact state of the model.

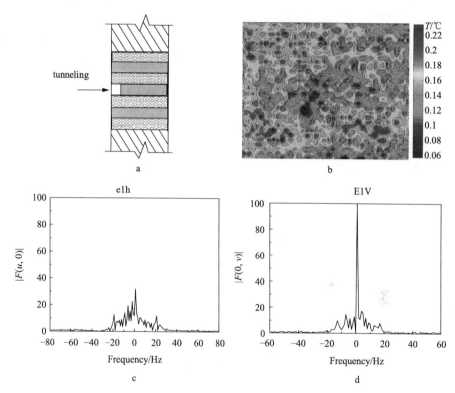

Figure 5.10 Figure set for the step 1 excavation E_1

a. 1/5 footage of the excavation; b. denoised thermograms; c. the horizontal spectrum $|F(u, 0)|$;
d. the vertical spectrum $|F(0, v)|$

In the thermal images E_1-E_4 (Figure 5.10b-Figure 5.13b), the EDZ exhibits the strip-formed and stratum-paralleled IRT during the early excavations when the face advanced shallow; but IRT distribution looks random, indicating the damage mechanism that the faulted beddings is dominant and damage propagates with a relatively small scale. In the thermal images E_5-E_6 (Figure 5.14b, Figure 5.15b), the EDZ are still strip-formed and stratum-paralleled but the IRT distributes in a localized configuration, representing the damage localization caused by the face advanced deep. Especially at E_6 (Figure 5.15b), the IRT is localized around the center of the excavation zone indicating the destruction of #1 RB.

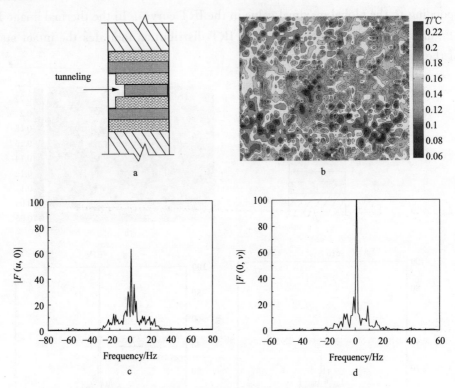

Figure 5.11 Figure set for the step 1 excavation E_2

a. 2/5 footage of the excavation; b. denoised thermograms; c. the horizontal spectrum $|F(u, 0)|$;
d. the vertical spectrum $|F(0, v)|$

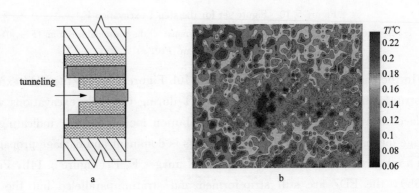

Chapter 5 Excavation in 45° strata

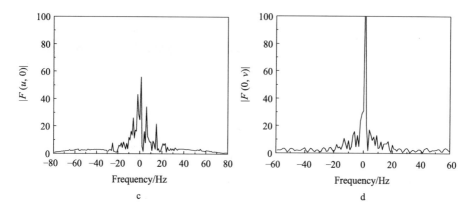

Figure 5.12 Figure set for the step 1 excavation E_3
a. 1/2 footage of the excavation; b. denoised thermograms; c. the horizontal spectrum $|F(u, 0)|$;
d. the vertical spectrum $|F(0, v)|$

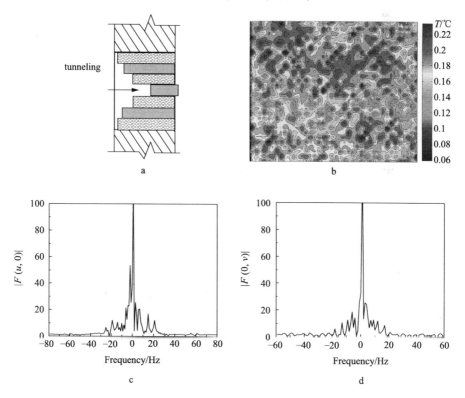

Figure 5.13 Figure set for the step 1 excavation E_4
a. 3/5 footage of the excavation; b. denoised thermograms; c. the horizontal spectrum $|F(u, 0)|$;
d. the vertical spectrum $|F(0, v)|$

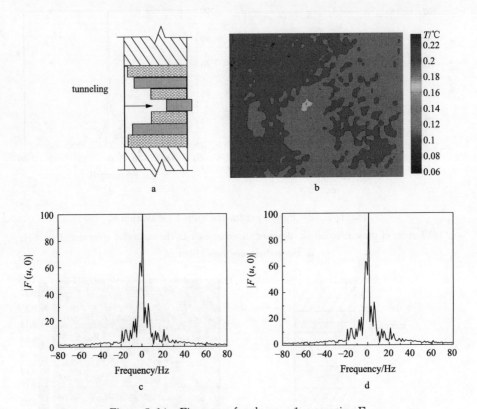

Figure 5.14　Figure set for the step 1 excavation E_5

a. 4/5 footage of the excavation; b. denoised thermograms; c. the horizontal spectrum $|F(u, 0)|$;
d. the vertical spectrum $|F(0, v)|$

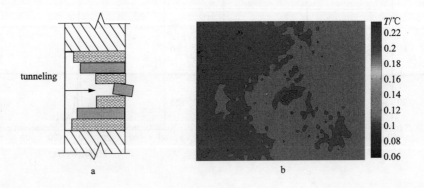

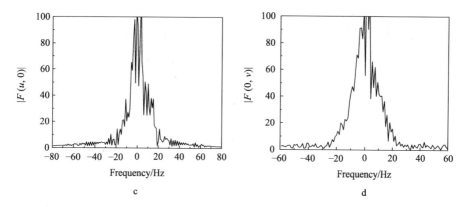

Figure 5.15 Figure set for the step 1 excavation E_6 (the critical point)
a. the rock block was destructed for the first time; b. denoised thermograms; c. the horizontal spectrum $|F(u, 0)|$; d. the vertical spectrum $|F(0, v)|$

From the directional spectra $|F(u)|$ and $|F(v)|$ (see each figure of c and d), we can evaluate the directional EDZ development at a specific instant. POI E_0 (Figure 5.9c、d): both the horizontal and vertical spectra were distributed with narrow bands and low amplitude at the same level representing the intact state of the rock mass. POIs E_1-E_3 (Figure 5.10-Figure 5.12c,d): the directional spectra have much higher amplitude in the vertical direction and the bandwidth become wider gradually, indicating the fact that when the face advanced shallow the nearby rockmass had undergone more damage along the vertical direction. POIs E_4-E_6 (Figure 5.13-Figure 5.15 c,d): the amplitude for the directional spectra has almost the same amplitude denoting that the EDZ developed in the horizontal and vertical directions at the same scale.

5.5.3 Comparison between the excavation in 0° and 45° inclined strata

For deepening our understanding in the EDZ with respect to the inclined rock layers, it is worthwhile to make comparison between the findings about the excavation in the horizontal strata in (He et al., 2010a) and this study. Figure 5.16 and Figure 5.17 shows the figure sets for the excavation steps E_1 and E_6 obtained in the experiment on the full-face excavation in the horizontal strata (He et al., 2010a, see chapter 7).

From Figure sets, we see that for the excavation step E_1 (Figure 5.16), the EDZ is scattered and for the excavation step E_6 (Figure 5.17), the EDZ is localized in a plastic from around the destructed #1 RB.

In contrast, it is seen from Figure sets for excavation in 45° inclined strata that, for the step 1 excavation E_1 (Figure 5.10), the EDZ is in a faulting bedding form and for step 6 excavation E_6 (Figure 5.15), the EDZ is also localized but with faulting damage along the same inclination with the strata.

As for the directional Fourier spectra, it is seen from the Figures 5.10c and d for the 45° inclined strata and from the Figures 5.16c and d for the horizontal strata, that no difference can be observed at the excavation step E_1, indicating the fact that the stress propagation in the horizontal and vertical direction is almost at the same intensity and the EDZ could be negligible.

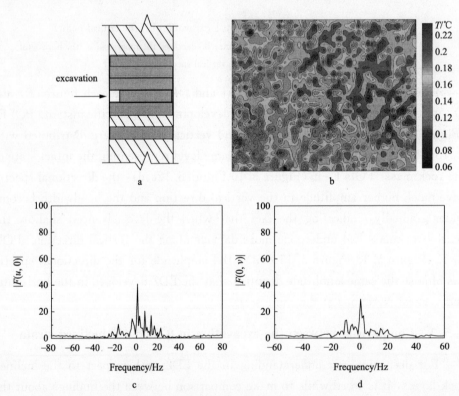

Figure 5.16 Figure set for the step 1 excavation E_1 in the horizontal strata (see chapter 7)
a. full-face excavation process for E_1 in the geologically horizontal strata; b. denoised thermograms;
c. the horizontal spectrum $|F(u, 0)|$; d. the vertical spectrum $|F(0, v)|$

At the critical point E_6, it is seen Figures 5.15c, d and Figures 5.17c, d, that the wave propagation patterns almost have the same both in the horizontal and vertical, but the spectra for the horizontal strata have wider bandwidths and higher amplitude, indicating the much more stronger rock damage had taken place in the horizontal strata.

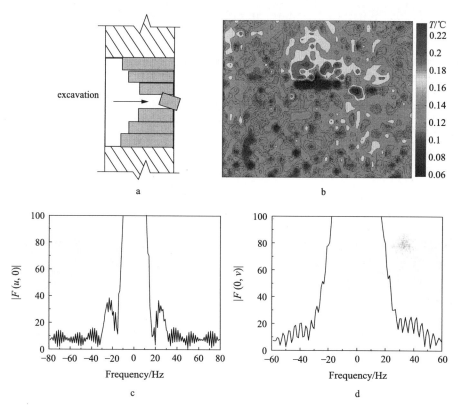

Figure 5.17 Figure set for the step 1 excavation E_6 in the horizontal strata, the critical point (see chapter 7)

a. full-face excavation process for E_6 in the geologically horizontal strata; b. denoised thermograms; c. the horizontal spectrum $|F(u, 0)|$; d. the vertical spectrum $|F(0, v)|$

5.5.4 Characterization of the staged excavation

Figure 5.18-Figure 5.25 show the figure sets for the staged excavation from P_0-P_7. The excavation diagram, thermal image and horizontal and vertical spectra are marked using letters a, b, c and d in the same order as those in the previous section. Rock responses over the staged excavation, depicted in Figures 5.18-Figure 5.25, can be understood as:

(1) Tunneling diagrams: the front view of the diagram shows the excavated space in every stage and 45° rotary sectional view of the diagram illustrates the

dynamical effects by removing each of the RBs. P_0 (Figure 5. 18a) shows the instant when a small passage was perforated in the #1 RB. P_1-P_7 (Figure 5. 19a-Figure 5. 25a) illustrate the status in every excavation stage.

(2) Thermal image P_0 (Figure 5. 18b): a small region of the IRT with deep blue-colored elsewhere on the image indicates the significant IRT drop induced by the unloading impact as a result of cutting through a small passage on the model for the first time, also shown in the IRT curve in Figure 5. 6 by the infrared image Po. From thermography P_1-P_7 (Figure 5. 19c-Figure 5. 25c) one can see the progressive development of EDZ around tunneling face. These EDZ were featured with damage localization in the nearby structures and faulted bedding damage along the same inclination as the rock strata. The scale of the localized damage was increased while the faulted bedding damage were getting fainter in contrast,

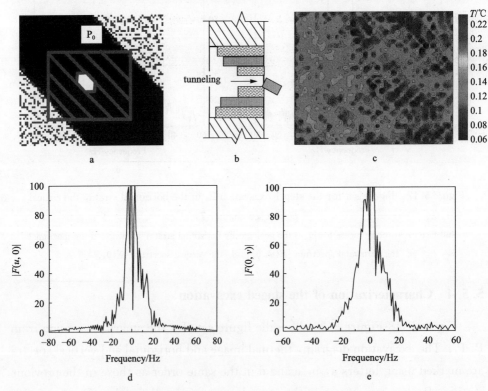

Figure 5. 18 Figure set for the excavation stage P_0
a. cutting through a small passage at a front view; b. cutting through a small passage at 45° rotary sectional view; c. denoised thermograms; d. horizontal spectrum $|F(u, 0)|$; e. vertical spectrum $|F(0, v)|$

indicating the fact that during the staged excavation, the faulting damage was getting lesser and the plastic damage around the face getting larger in scale as the roadway tunnel excavation was getting to be finished.

(3) The directional spectra $|F(u,0)|$ and $|F(0,v)|$: from the figure sets c and d in Figure 5. 19-Figure 5. 25 we can see that from POIs P_1-P_7, the horizontal and the vertical spectra have almost the same bandwidth and amplitude, indicating the fact that the excavation induced stress wave propagates at the same level in the horizontal and vertical direction; the bandwidth for the two directional spectra varied from narrow to wider and finally to narrower with the excavation approaching to an end, indicating the excavating damage level undergone inside the surrounding rock masses over the staged excavation.

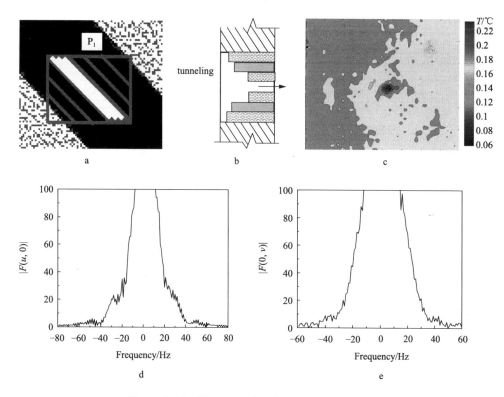

Figure 5. 19　Figure set for the excavation stage P_1
a. removing #1 RB at a front view; b. removing #1 RB at 45° rotary sectional view; c. denoised thermograms; d. horizontal spectrum $|F(u, 0)|$; e. vertical spectrum $|F(0, v)|$

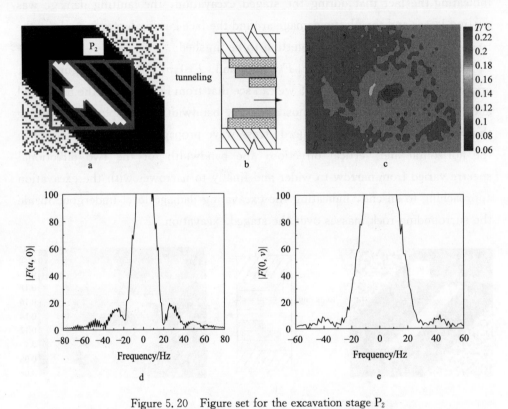

Figure 5.20 Figure set for the excavation stage P_2

a. removing #2 RB at a front view; b. removing #2 RB at 45° rotary sectional view; c. denoised thermograms; d. horizontal spectrum $|F(u, 0)|$; e. vertical spectrum $|F(0, v)|$

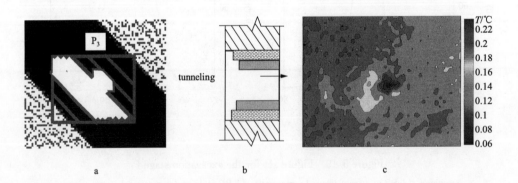

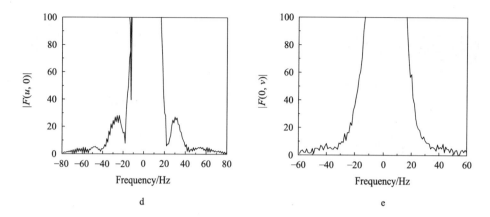

Figure 5.21 Figure set for the excavation stage P_3
a. removing #3 RB at a front view; b. removing #3 RB at 45° rotary sectional view; c. denoised thermograms; d. horizontal spectrum $|F(u, 0)|$; e. vertical spectrum $|F(0, v)|$

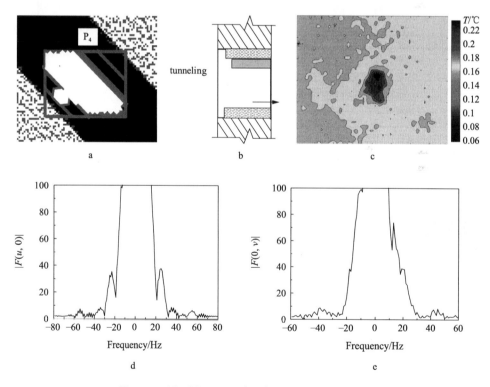

Figure 5.22 Figure set for the excavation stage P_4
a. removing #4 RB at a front view; b. removing #4 RB at 45° rotary sectional view; c. denoised thermograms; d. horizontal spectrum $|F(u, 0)|$; e. vertical spectrum $|F(0, v)|$

Infrared Thermography for Geomechanical Model Test

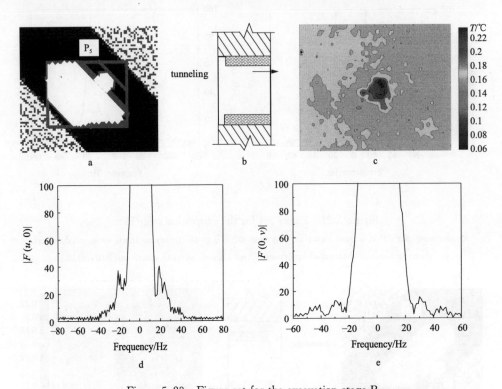

Figure 5.23 Figure set for the excavation stage P_5

a. removing #5 RB at a front view; b. removing #5 RB at 45° rotary sectional view; c. denoised thermograms; d. horizontal spectrum $|F(u, 0)|$; e. vertical spectrum $|F(0, v)|$

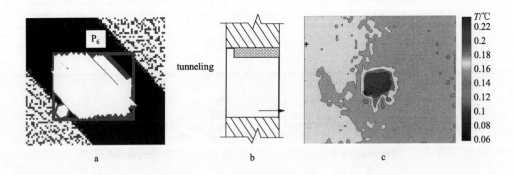

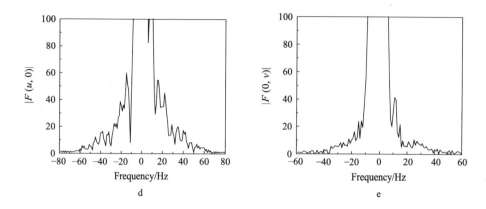

Figure 5.24 Figure set for the excavation stage P_6

a. removing #6 RB at a front view; b. removing #6 RB at 45° rotary sectional view; c. denoised thermograms; d. horizontal spectrum $|F(u, 0)|$; e. vertical spectrum $|F(0, v)|$

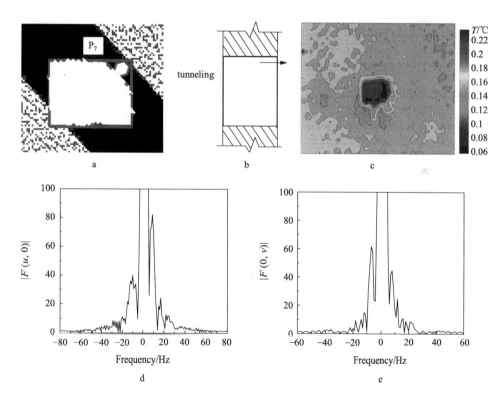

Figure 5.25 Figure set for the excavation stage P_7

a. removing #7 RB at a front view; b. removing #7 RB at 45° rotary sectional view; c. denoised thermograms; d. horizontal spectrum $|F(u, 0)|$; e. vertical spectrum $|F(0, v)|$

The significant difference for the EDZ over the staged excavation between the horizontal strata (He et al., 2010a) and this study lies in the spatial configurations. That is, for the horizontal strata, the EDZ was in the plastic form and distributed around the tunneling face [for detail see chapter 7 and literature by He et al. (2010a)]; for the 45° inclined rock strata, however, the faulted bedding damage was seen over the whole staged excavation processes. This observation revealed the dependency of the EDZ upon the inclinations for the jointed rockmasses.

5.5.5 Summary

IRT was used as a measure in characterizing the rock behaviour during the excavation processes. Over the full-face excavation, the IRT curve has a piecewise linear increasing pattern between the jump discontinuity points which was caused by the sliding effect at a macroscopic scale; and during the staged excavation, the IRT curve has a fluctuating pattern due to the loading and unloading impacts by removing the rock blocks. As a result, the IRT characterizes the full-face excavation and the staged excavation as "multi-linearity" and nonlinearity processes respectively, in an energy dissipation perspective within the infrared radiation regime.

Infrared thermography series characterize the dynamical process of the EDZ development by their spatial IRT distributions over the roadway excavation processes. During the full-face excavation, the EDZ is in the form disturbed belt parallel to the rock strata indicating faulted bedding dominant damage mechanism. During the staged excavation, the EDZ is in the form of localization around the tunneling face with the faulted bedding damage seen over the staged excavation process.

The directional propagation of the excavation induced stress wave was characterized by horizontal and vertical Fourier spectra obtained by performing DFT on the data sets by resampling on the infrared image matrix along the horizontal and vertical symmetry axes respectively. More stress waves with higher level amplitude was found in the early excavations during the full-face excavation; and the stress wave components were found almost getting to be uniform as the tunneling face advancing deep and over the whole staged excavation process.

References

Barla M. 2008. Numerical simulation of the swelling behaviour around tunnels based on special triaxial tests. Tunnelling and Underground Space Technology, 23: 618-628.

Brady B T, Rowell G A. 1986. Laboratory investigation for the electrodynamics of rock fracture. Nature, 321: 488-492.

Cai M, Kaiser P K, Morioka H, et al. 2007. FLAC/PFC coupled numerical simulation of AE in large-scale underground excavations. International Journal of Rock Mechanics and Mining Sciences, 44: 550-564.

Cai M. 2008. Influence of stress path on tunnel excavation response-Numerical tool selection and modeling strategy. Tunnelling and Underground Space Technology, 23: 618-628.

Chambon P, Corte J F. 1994. Shallow tunnels in cohesionless soil : stability of tunnel face. J. Geotech. Eng. , 120(7): 1148-1165.

Connolly M, Copley D. 1990. Thermographic inspection of composite material. Mater Evaluation, 48(12): 1461-1463.

Cui C, Deng M, Geng N. 1993. Study on the features of spectrum radiation of rocks under different load. Chinese Sci. Bulletin, 38(6): 538-541.

Geng N G, Yu P, Deng M D, et al. 1998. The simulated experimental studies on cause of thermal infrared precursor of earthquake. Earthquake, 18: 83-88.

Grinzato E, Marinetti S, Bison P G, et al. 2004. Comparison of ultrasonic velocity and infrared thermography for the characterization of stones. Infrared Phys. Tech. , 46: 63-68.

Harwood N, Cummings W M. 1991. Thermoelastic Stress Analysis. Bristol: IOP Publishing Ltd.

He M C. 2006. Rock mechanics and hazard control in deep mining engineering in China. Rock Mechanics in Underground Construction. In: Rock Mechanics in Underground Construction, Proceedings, ISRM International Symposium 2006 4th Asian Rock Mechanics Symposium: 14-31.

He M C, Gong W L, Li D J, et al. 2009. Physical modeling of failure process of the excavation in horizontal strata based on IR thermography. Min. Sci. Tech. , 19(6): 689-698.

He M C, Gong W L, Zhai H M, et al. 2010a. Physical modeling of deep ground excavation in geologically horizontal strata based on infrared thermography. Tunnelling and Underground Space Technology, 25: 366-376.

He M C, Jia X N, Gong W L, et al. 2010b. Physical modeling of an underground roadway excavation vertically stratified rock using infrared thermography. International Journal of Rock Mechanics and Mining Sciences, doi:10. 1016/j. ijrmms. 2010. 06. 020.

He M C, Miao J L, Feng J L. 2010c. Rock burst process of limestone and its acoustic emission characteristic under true-triaxial unloading conditions. International Journal of Rock Mechanics and Mining Sciences, 47: 286-298.

He M C, Lu X J, Jing H H. 2002. Characters of surrounding rock mass in deep engineering and its non-linear dynamic-mechanical design concept. Chin. J. Rock Mech. Eng., 21(8): 1215-1224.

He M C, Yong X J, Sun X M. 2006. Chinese Coal Mine Soft Rock-study on Characteristics of the Clay Minerals. Beijing: Coal Industry Press.

Jia P, Tang C A. 2008. Numerical study on failure mechanism of tunnel in jointed rock mass. Tunnelling and Underground Space Technology, 23: 500-507.

Kamata G, Masimo H. 2003. Centrifuge model test of tunnel face reinforcement by bolting. Tunnelling and Underground Space Technology, 18(2): 205.

Lee Y Z, Schubert W. 2008. Determination of the length for tunnel excavation in weak rock. Tunnelling and Underground Space Technology, 23: 221-231.

Li S J, Yu H, Liu Y X, et al. 2008. Results from in situ monitoring of displacement, bolt load, and disturbed zone of a power house cavern during excavation process. International Journal of Rock Mechanics and Mining Sciences, 45: 1519-1525.

Luong M P. 1990. Infrared thermovision of damage processes in concrete and rock. Eng. Fracture Mech., 35(1/2/3): 291-310.

Long M P. 1995. Infrared thermographic scanning of fatigue in metals. Nuclear Eng. Design, 158: 363-376.

Majewska Z, Mortimer Z M. 1998. Fractal description of acoustic emission produced in systems: coal-gas and gas-water. J. Acoust. Emiss, 16:1-4.

Meguid M A, Saada O, Nunes M A, et al. 2008. Physical modeling of tunnels in soft ground: A review. Tunnelling and Underground Space Technology, 23: 185-198.

Park S H, Adachi T, Kimura M, et al. 1999. Trap door test using aluminum blocks, In: Proceedings of the 29th Symposium of Rock Mechanics. J. S. C. E. : 106-111.

Pastor M L, Balandraud X, Grédiac M, et al. 2008. Applying infrared thermography to study the heating of 2024-T3 aluminium specimens under fatigue loading. Infrared Phys. Tech., 51: 505-515.

Pinsky M A. 2003. Introduction to Fourier Analysis and Wavelets. Beijing: Chin. Machine Press.

Qiang Z, Xu X, Ning C. 1990. Abnormal infrared thermal of satellite: Forewarning of earthquake. Chinese Sci. Bulletin, 35(17): 1324-1327.

Read RS. 2004. 20 years of excavation response studies at AECL's Underground Research Laboratory. International Journal of Rock Mechanics and Mining Sciences, 41: 1251-1275.

Sharma J S, Bolton M D, Boyle R E. 2001. A new technique for simulation of tunnel excavation in a centrifuge. Geotech. Testing J., 24(4): 343-349.

Shiotani T. 2006. Evaluation of long-term stability for rock slope by means of acoustic emission technique. NDT & E int., 39:217-228.

Shi W Z, Wu Y H, Wu L X. 2007. Quantitative analysis of the projectile impact on rock using infrared thermography. Int. J. Impact Eng., 34: 990-1002.

Sitharam T G, Latha G M. 2002. Simulation of excavations in jointed rock masses using a practical equivalent continuum approach. International Journal of Rock Mechanics and Mining Sciences, 2002; 39: 517-525.

Steinberger R, Valadas L T I, Ladstätter E, et al. 2006. Infrared thermographic techniques for non-destructive damage characterization of carbon fiber reinforced polymers during tensile fatigue testing. Int. J. Fatigue, 28: 1340-1347.

Tanaka T, Sakai T. 1993. Progressive failure and scale effect of trapdoor problem with granular materials. Soils Foundations, 33(1): 11-22.

Tang Y G, Kung G T C. 2009. Application of nonlinear optimization technique to back analyses of deep excavation. Compt. Geotech, 36: 276-290.

Wu B R, Lee C J. 2003. Ground movement and collapse mechanisms induced by tunneling in clayey soil. Int. J. Phys. Modeling Geotech. , 3(4): 13-27.

Wu L X, Liu S J, Wu Y H, et al. 2002. Technical Note: Changes in infrared radiation with rock deformation. International Journal of Rock Mechanics and Mining Sciences, 39: 825-831.

Wu L X, Liu S J, Wu Y H, et al. 2006a. Precursors for rock fracturing and failure-Part I: IRR image abnormalities. International Journal of Rock Mechanics and Mining Sciences, 43: 473-482.

Wu L X, Liu S J, Wu Y H, et al. 2006b. Precursors for rock fracturing and failure-Part II: IRR T-Curve abnormalities. International Journal of Rock Mechanics and Mining Sciences, 43: 483-493.

Wu L X, Wang J Z. 1998. Technical note: Infrared radiation features of coal and rocks under loading. International Journal of Rock Mechanics and Mining Sciences, 35(7): 969-976.

Wu L X, Wu Y H, Liu S J, et al. 2004. Technical Note: Infrared radiation of rock impacted at low velocity. International Journal of Rock Mechanics and Mining Sciences, 41: 321-327.

Zhi Y, Cui C, Zhang J. 1996. Application of infrared imaging system to the basic remote sensing experiment on rock mechanics. Remote Sens. Environ. Chin. , 11(3): 161-167.

Chapter 6 Excavation in horizontal strata

6.1 Introduction

Comprehensive understanding of rock response to the roadway excavation in deep ground has long been the interests for the community of geomechanics, for the nonlinearity and time-dependent natures of the rocks at great depth (He, 2006).

Both in situ and laboratory experiments on the excavation problems have been performed with different methods. For instance, Tang and Kung (2009) conducted field observations incorporated with finite element code for the back analysis of excavation-induced wall deflection. Li et al. (2008) carried out in-situ monitoring for the assessment of displacement, bolt load, and disturbed zone of a powerhouse cavern during excavation process.

Field experiments were conducted at Underground Research Laboratory to investigate the formation of rock damage around tunnels, and to assess the factors that influence the stability of excavations, by recording displacements, shaft convergence, stress changes, and microseismic events in the rock as excavation processed (Read, 2004). The characteristics of the EDZ (Excavation Damaged Zone) developed during the construction of the underground research tunnel were investigated through in situ tests and laboratory core testing (Kwon et al., 2009), etc.

Nevertheless, field investigations and full-scale experiments are very expensive, difficult to run and hard to repeat. As a result, experimental tests with a reduced physical model have been the most effective means to provide sufficient and reliable information for the rock response to the underground excavations, e.g. the failure mechanism of the tunnel face (Chambon and Corte, 1994), and stability of the face and unsupported span under tunnel excavation in weak rock (Lee and Schubert, 2008).

Over decades, various methods with a reduced physical model have been proposed for simulating the tunnel excavations, including the trap door model tests (Tanaka and Sakai, 1993), the polystyrene foam and organic solvent experiments (Sharma et al., 2001), the rigid tube with flexible or movable face tests (Kamata and Masimo, 2003), pressurized air in a rubber bag of negligible strength methods (Wu and Lee, 2003), and a miniature tunnel boring machine (TBM) techniques (Nomoto et al., 1999; Meguid et al., 2008), etc.

The limitations of the small scale model concern the inability of faithful simulation of the in-situ stresses, and the inconsistencies in scaling factors for different variables (e. g. length, inertia force, creep, etc). Therefore, new physical modeling approaches with a large-scale physical model, incorporated with the state-of-the art measuring approaches, need to be introduced into the roadway excavation tests in order to simulate realistically the structural and mechanical responses of rock masses under tunneling.

Infrared (IR) thermography, as a non-destructive, remote sensing technique, has been widely used in detection of the onset of unstable crack propagation and/or flaw coalescence for concrete and rock, based on the fact that the heat generation is caused by the intrinsic dissipation due to elasticity and inelasticity of the material under external loading (Toubal et al., 2006).

Infrared imaging devices offer remote, accurate, and fast assessment of temperature rise due to a fault inside materials, and IR signals are readily converted to visual display to provide advance warning of potential danger areas, i. e., thermographic imaging can detect fault invisible to the naked eye (Luong, 1990). The abnormities of the thermal image, the features of the IR temperature, and the stress levels as an advanced warning signature preceding the structural failure were investigated for coal rock and sandstone subjected to some given loading schemes (Wu and Wang, 1998; Meola, 2007).

During deformation and failure process of rocks, thermography records temperature variations on the surface in view and displays it as infrared images with false colors, where a high temperature (in warm color) denotes shear fracturing from the frictional effects; low temperature (in cool color) represents the tensile fractures indicating a permanent plastic damage (Connolly and Copley, 1990; Steinberger et al., 2006; Wu et al., 2006).

The chapter reports the findings from the geomechanical model test on roadway excavation in the horizontally inclined stratified rocks. The PFESA (Physically

Finite Elemental Slab Assemblage) method was employed for construction of the large-scale physical model. The roadway excavation using drill and blasting method without support was carried out over the two phases, i. e. , the full-face excavation and staged excavation.

Infrared (IR) thermography, incorporated with such image processing procedures as data statistics, noise removal and 2-dimentional DFT (Discrete Fourier Transformation) for extracting features from the resulting thermographies, were employed to record and capture the progressive development of the excavation response for the PFESA model under the roadway tunneling impact.

6.2 Experiment

6.2.1 Geomechanical model construction

The method for conducting the dimensional analysis and constructing the geomechanical models have been discussed in chapter 4, 5 and 6 in this book. Same procedures were implemented for construction of the geomechanical model consisting the horizontal layers of sandstone, mudstone and coal seam. Major parameters for describing the model are summarized in the following.

Dimensions of the geomechanical model were designed based on the dimensional analysis of the prototype in QISHAN underground coal mine based on the similarity principles (see chapter 4). The model is 1.6 mm×1.6 mm×0.4 mm (length× height× thickness). The roadway excavation zone was centered on the model within the Stratum 4 (coal seam), measuring 0.25 m×0.2 m at the cross section and 0.4 m for the corridor length (throughout the thickness of the model).

The constructed geomechanical model is shown schematically in Figure 6.1, where Figure 6.1a is the schematic drawing, Figure 6.1b is the photograph taken at the laboratory, and Figure 6.1c is the rock reality of the deep underground roadway simulated by the PFESA model. According to the geological histogram chart of the QI SHAN coal mine, the PFESA model was assembled with nine strata. The strata are numbered by 1-9. The simulated rock types and the geometrical parameters for the strata are reported in the geological section (see Table 6.1)

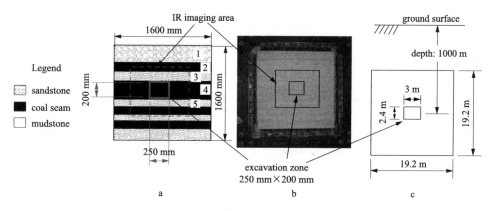

Figure 6.1　The tested PFESA model
a. schematic of the model; b. photograph of the model; c. rock reality of the deep underground cavern simulated by the PFESA model

Table 6.1　Geological section, material and structural parameters of the PFESA model

Geological section	Stratum No.	Rock types	Layer thickness/mm	Layer numbers
	1	sandstone	440	14
	2	coal seam	140	10
	3	mudstone	120	7
	4	coal seam	250	17
	5	mudstone	150	5
	6	coal seam	60	4
	7	mudstone	140	5
	8	coal seam	60	3
	9	mudstone	240	8

6.2.2　Testing procedure

1. Excavation scheme

The earthwork for roadway excavation is a a cubic volume measuring 250 mm × 200 mm × 400 mm, located on the Stratum 4 (coal seam), as seen in Figure 6.1. Like the excavation plan in chapter 4 and 5, the whole tunneling volume was divided into seven portions, referred to as "rock block" (RB). The excavation plan is shown in Figure 6.2, where Figure 6.2a is the front face of the excavation zone and 6.2b is the cross section of the roadway corridor.

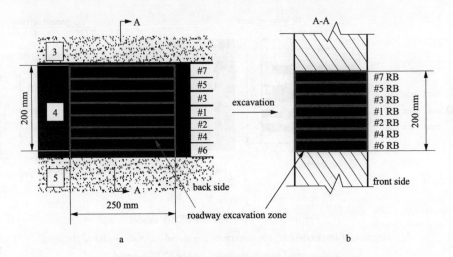

Figure 6.2 Roadway excavation plan
a. front face of the excavation zone on the model; b. sectional view of the excavation zone

The RBs were indexed #1 to #7 RB in accordance with the excavation sequence. The roadway excavation was performed by two phases: full-face excavation, i.e., tunneling on the #1 RB and the adjacent rocks until a small passage is cut through; and the staged excavation, i.e., removing one RB at each tunneling stage.

Rock behaviors in responding to the volume of earthwork (referred to as footage thereafter) during the full face excavation and removal of the key slabs (i.e., RBs) during the staged excavation were investigated in terms of the EDZ development. The excavation started at the back side of the model, and went through to the front face, with a chisel and a hammer as the tunneling tools.

2. Excavation method

The excavation of the roadway tunnel without support was started at the back side of the model, and went through to the front face, with a chisel and a hammer as the tunneling tools, simulating the tunnel excavation by drill and blast method. Detailed description of the excavation method can be found in chapter 4 (see Figure 4.6).

3. Boundary condition

During the excavation, the vertical load and lateral load were applied on the model (simulated as quasi-2D plane strain state) by using frame (described in chapter 4), as shown schematically in Figure 6.3. In order to reproduce the unbalanced stress state on the boundaries of the model, the lateral pressure coefficient $\lambda = \sigma_1/\sigma_2 = 0.3$ was evaluated from the in-situ investigation (σ_1 and σ_2 are the vertical and lateral stresses respectively). According to the coefficient λ, the vertical and lateral stresses were chosen as 1.0 and 0.3 MPa.

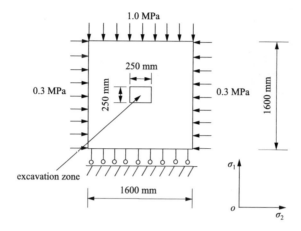

Figure 6.3 Simulated boundary stress condition over the roadway excavation

6.3 Infrared detection

6.3.1 Infrared thermography

The infrared thermography TVS-8100MK II was employed for capturing the infrared radiation emittances during the roadway excavation. During the test, the image acquisition frequency was set as one frame every four seconds for the IR thermography, and the detection distance between the infrared camera and the model face was kept at 1333 mm, in order to have a viewed area of 400 mm×267 mm, in coverage of the roadway cross section. The experimental setup is shown in Figure 6.4; Figure 6.4a is the photograph and Figure 6.4b the schematic.

Twenty four hours prior to the testing, all the instruments were placed in the same room with the model, so that the detected IR temperatures are the temperature variations induced by external loading. The 400 mm×400 mm frame of red colored plastic membrane was attached with glue to the front face of the model, symmetric to the cross section of the roadway, indicating the infrared imaging area, as seen in Figure 6.4a. The infrared camera and video camera were

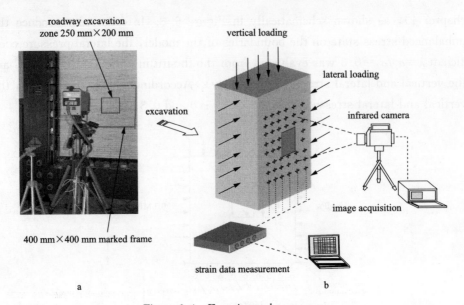

Figure 6.4 Experimental test setup
a. photograph of the IR camera and the excavation area in view; b. the schematic diagram for the testing system

both targeted at the model's front face, and the excavation operations commenced at the rear face, recording the thermal image and displaying them for observation of the instant change and progressive development of the excavation damage on-the spot.

6.3.2 Image processing

1. Image sharpening algorithm

From an information comprehension point of view, the useful information implied by an image lies in the locations where the gray level changes sharply. To be specific, for an IR image, the boundaries or the edges encompassing the high or low infrared temperature regions are the most interested objects to be separated out. The first-order derivatives are widely used in the segmentation of these interested objects, known as *"image sharpening"*.

The mathematical nature of the first-order derivative in the image processing is the absolute value of the gradient, defined by

$$G[f(x,y)] = \max\{|\,\mathrm{grad}[g(x,y)]\,|\} = [(\partial g/\partial x)^2 + (\partial g/\partial y)^2]^{1/2}$$

(6.1)

where, $g(x,y)$ is the pixel matrix computed by Eq. (6.1); grad denotes the gradient operator in the IPT; $f(x, y)$ is the denoised IR image with the same dimensions as the input image $f'(x,y)$. It is to be noted that $f(x,y)$ in this study has a double meaning: i. e., denoting the IR image itself or the image matrix, depending on the context in which it is used.

The denoised IR image $f(x,y)$ was used for visualization and description of initiation, propagation and coalescence of the excavation damage. In this study, the image segmentation was performed by using the first-order derivative based Canny operator, for removing the coherent noises and sharpening the IR images. Detailed introduction of the concerning theories for the image noise processing is available in the current publications, such as literature (Gonzalez et al., 2005).

2. Pretreatment of the thermal image algorithm

The following two pretreatment algorithms for the proper image processing are required (for details see chapter 4):

(1) subtraction of the first frame from the follow-up IR images for eliminating the background radiation noises;

(2) performing the median filter for suppressing the impulsive noise.

The above image processing operations were realized in the MATLAB 8.0 platform based on the Image Processing Toolbox (IPT) functions in the Matlab macro code.

3. Normalized energy release index

The *energy release index* is defined as the normalized mean value of the thermal image matrix, expressed as "*normalized* ⟨IRT⟩ or ⟨IRT⟩ or IRT", depending on the context of the documentation, for the sake of simplicity. The *normalized* IRT was used for characterization the overall rock responses to the excavation in terms of the energy release emitted from the stresses rocks. By employing Eq. (5.3) and Eq. (5.4), the normalized ⟨IRT⟩ can be calculated using the thermal sequence acquired in this test. Detailed introduction is given in sub-section 5.4.2 in chapter 5.

6.3.3 Fourier transform of the thermal image

In underground excavation, the structure response is stress path dependent for nonlinear materials, and a sudden excavation impact creates large unbalance forces right at the excavation boundary and the unbalanced forces need to be released, which will generate dynamic stress waves (Ruistuen and Teufel, 1996; Kaiser et al., 2001; Cai et al., 2002). The excavation induced stress wave, consisting of compressive and tensile wave, will travel in the rock layers and reflect at the interfaces of the layers, provoking variations of the IR signals on the viewed surface.

Stress wave propagation can be described in the frequency domain by a 2-dimensional (2-D) Fourier transformation of the IR images. The 2-D Fourier spectra of the IR images were achieved by the 2-D DFT (discrete Fourier transformation) algorithm, which is defined by (Mark, 2003):

$$F(u,v) = \mathrm{DFT}[f(i,j)] = \sum_{i=0}^{M-1}\sum_{j=0}^{N-1} f(i,j)\exp\left[-2\pi i\left(\frac{ui}{M}+\frac{vj}{N}\right)\right] \quad (6.2)$$

where, u and v are the horizontal and vertical frequency variables respectively, $u=1,2,\cdots,M$; $v=1,2,\cdots,N$; $F(u, v)$ is the 2-D Fourier spectrum, also referred to as the spatial spectrum. The amplitude spectrum $F(u, v)$ is utilized in the subsequent analysis to characterize the excavation induced stress waves in the frequency domain.

The reciprocal value of the variables is the wavelength λ for $|F(u, v)|$, i.e., $\lambda = 1/\sqrt{u^2+v^2}$ [the dimensions for these variables are: λ (m); u(Hz); v (Hz)]. The scope of the frequency variables (also termed as "frequency band" or "band") and the related amplitude, $|F(u, v)|$, describes the propagation of the stress waves due to the fracturing process of the rock mass. The terms of "wave components/or wave" and "spectrum components/or spectrum" were identically or alternatively utilized to describe the excavation induced stress waves in the following context.

6.3.4 Enhancement of the thermal image

The thermography works at the middle infrared band from 3.6 to 4.6 μm, capable of detecting very small changes in temperature radiation. The IR camera works at the IR spectral regions (detection wavelength 3.6 to 4.6 μm for the type used in our test), capable of detecting very small changes in IR radiation. Infrared thermography, incorporated with the noise removal algorithm proposed in this paper, have enabled us to develop better pictures of the progressive development of the excavation damaged zone (EDZ), which would otherwise be undetectable and observable, especially at the early phase of the roadway excavation.

Figure 6.5 presents an example of the image processing work for exhibiting the instantaneous excavation damage at 3/5 footage (marked by E_4). Figure 6.5a is the photograph taken at the 3/5 footage of the tunneling work during the full-face excavation. The slightly prominent portion on the excavation zone was the sliding of the slabs with a small displacement due to the tunneling on the #1 RB. It is noted that no in-plane deformation could be observed or measured at macroscopic scale as a result of the sliding displacement perpendicular to the surface of the model.

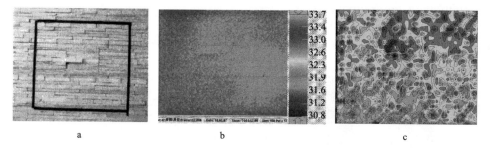

Figure 6.5 An example of effect of noise removal procedure for exhibiting the excavation damage at 3/5 footage (marked by E4) during phase 1 excavation

a. photograph; b. original IR image; c. same IR image after denoising

Figure 6.5b is the raw thermal image on which we can only see a little difference of the contrast on the location of the sliding slabs. Figure 6.5c is the same frame of the thermal image after denoising by using the algorithm introduced in the subsection 6.5 in this chapter. It is seen that the permanent damage by the sliding of the slabs was depicted by a rectangle-shaped area in deep blue color (low IRT caused by the loosening of the rock beddings); and high level IRT distribution (caused by the frictional effect of the cracked materials) distributes around the sliding slabs, indicating the in-plane rock damage at micro-and meso-scales.

By using the denoised IR images, we evidenced the damage and failure mechanisms of the stratified soft strata during roadway excavation processes. It should be stated that in the following context, the configurations and scales of the high level IRT were used for the description of the EDZ at the specific time instant due to the initiation and propagation of the microscopic fractures in terms of the "IRT distribution", and the low level IRT (deep blue colored area) in larger scale represents a permanent or plastic damage due to the material failure, coalescence of the macroscopic fractures, as well as the excavated space in the model.

6.3.5 Spectral analysis

The excavation induced stress waves at excavation step E_4 can be characterized in frequency domain by the 2-D Fourier spectrum computed from the thermographies by using DFT introduced in the sub-section 6.6. Figure 6.6a is the 2-D Fourier spectrum $|F(u, v)|$, the origin of the coordinates (0, 0) indicates low frequency components; the bright regions represent distribution of the wave components; the level of brightness is directly proportional to the level of the amplitude of the components.

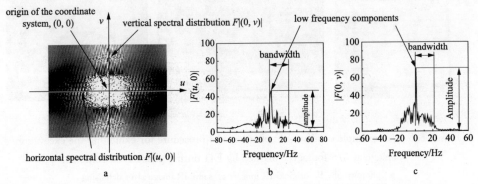

Figure 6.6 2-D Fourier spectral characterization of the excavation damage at 3/5 footage (marked by E_4) during phase 1 excavation

a. 2-D Fourier spectrum; b. horizontal spectral distribution $F|(u, 0)|$; c. vertical spectral distribution $F|(0, v)|$

From Figure 6.6a we can see a square-formed area centered on the spatial frequency plane having high-brightness, representing the principal components distribution; meanwhile, the dispersed regions with lesser brightness far away from the plane center, representing the very-high-frequency components with small amplitudes, which might be the information of latent danger preceding a material failure.

Figure 6.6b and c are the sectional spectral distributions of the 2-D spectrum along the horizontal axis, i.e., $|F(u, 0)|$, and vertical axis, i.e., $|F(0, v)|$, respectively. The sectional spectral distributions describe the excavation responses in the following aspects:

(1) the energy dissipation levels by the amplitude of the wave components, i.e., the energy dissipation level is directly proportional to the amplitude of the

low frequency component;

(2) the scales of the fractures of cracking materials are represented by the bandwidth (or band) for the high frequency components with a relatively higher amplitudes, i. e., wider bands represent the fractures with a wide range of scales (defect coalescence or damage localization);

(3) anisotropic nature of the stress wave in the plastic materials by comparing the amplitudes or bands of the wave components in the horizontal (parallel to the rock layers) and vertical (perpendicular to the rock layers) directions, i. e., broader band or higher amplitude denote higher level of the excavation damage undergone by the stratified rock strata.

6.4 Results and discussions

6.4.1 Overall thermal response

The characteristics of IRT are presented in Figure 6.7, and Figure 6.8 and Figure 6.9 are the excavation diagrams for helping the readers to understand the energy release index plotted in Figure 6.7. Figure 6.7 shows the temporal evolution of the IRT against the roadway excavation processes, and Figure 6.8 depicts the tunneling face advancement over the full-face excavation, and Figure 6.9 illustrates the tunneling stages over the staged excavation.

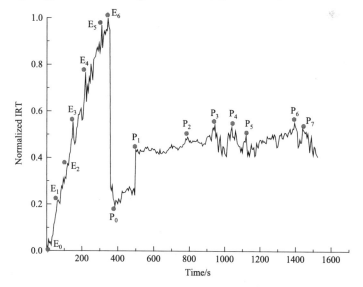

Figure 6.7 Profile of the measured IRT against the excavation advancement

According to the typical rock responses over the two phases of the excavation, some specific points (points of interest, POI) were selected on the IRT profile marked by the capital letters E_0 to E_6 (excavation steps in the full-face excavation) and P_0 to P_8 (excavation stages over the staged excavation) respectively (also see Figure 6.8). E_0 to E_6 are in accordance with the full-face excavation, i. e. , the tunneling was limited on the central part of the excavation area (#1 RB and neighboring area). E_0 is the beginning of the excavation, E_1 to E_5 with respect to the excavation advancement from step 1 to step 5, and E_6 with respect to the moment at which a small passage was about to cut through (destruction of the #1 RB, the critical point).

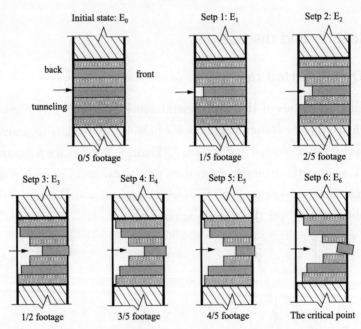

Figure 6.8 Diagram for the full-face (phase 1) excavation process at a sectional view

The IRT profile attains its maximal value at the excavation step E_6, before and after which the time-marching schemes of the IRT are distinct. Thus the POI E_6, representing the maximum IRT, i. e. , T_{max}, was denominated as "critical point". The IRT profile increases linearly during the phase 1 excavation process. As stated above, IRT measures the overall rock response to the roadway tunneling. So the rock behavior represented by the time-marching scheme of the IRT profile (marked by E_0 to E_6) can be treated as a linear process from an energy dissipation point of view.

After the critical point E_6 the IRT profile fluctuates with multiple local peaks. As a result, the rock behavior, marked by P_0 to P_7, can be treated as a non-linear process in the IR radiation regime (see Figure 6.7 and Figure 6.9). By calibration with the excavation advancement during the staged excavation, most of the major local peaks on the IRT profile correspond to the impact loading by removing the RBs (rock blocks). These specific points on the IRT curve were marked by P_0 to P_7.

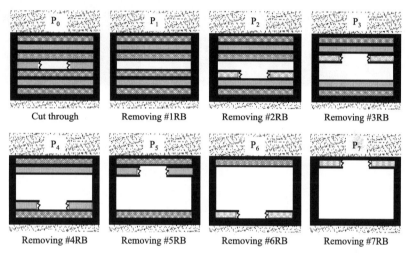

Figure 6.9 Diagram for the staged (phase 2) excavation process at a front view

Point P_0 indicates a sharp IRT drop immediately after the critical point E_6, due to the unloading impact by the perforation of a small passage in the #1 RB for the first time. Therefore the #1 RB can be regarded as the key slabs by the fact that destruction of the #1 RB (labeled by E_6 on the IRT curve) alters the rock behavior represented by the time-marching scheme of IRT in the IR radiation domain. During the phase 2 excavation (see Figure 6.9), only one RB was being excavated at each stage. Removal of the RBs caused a quasi-cyclic loading and unloading impact, resulting in the multiple local peaks on the IRT profile marked by P_1 to P_7, as seen in Figure 6.7.

6.4.2 Characterization of the full-face excavation

Figures 6.10 to Figure 6.16 show the figure sets for the excavation steps from E_0 to E_6 during the full-face excavation. In each of the figure set, the first figure is the diagram of the excavation steps (marked by a); the second the thermal image (marked by b); the 2-D Fourier spectrum $|F(u, v)|$ (marked by c); the horizontal spectrum along the horizontal axis u $|F(u, 0)|$ (marked by d and the vertical spectrum along the vertical axis v $|F(0, v)|$ (marked by e), respectively.

Figure 6.10a shows the initial state of the roadway excavation denoted by E_0, when the excavation was about to begin.

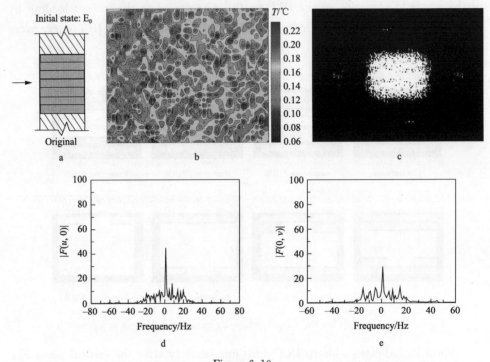

Figure 6.10

a. intact state of the model; b. IR images after denoising; c. 2-D spectral distribution $|F(u, v)|$; d. horizontal spectral distribution $|F(u, 0)|$; e. vertical spectral distribution $|F(0, v)|$

Figure 6.11a (E_1) and Figure 6.12a (E_2) show the tunneling advancement at the 1/5 and 2/5 footages of the earthwork, there was no sliding between the rock layers as the face advances shallow away from the surface of the model at the back side.

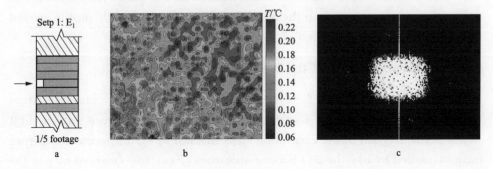

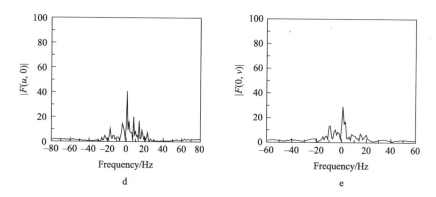

Figure 6.11　Figure set for step 1 excavation

a. excavation diagram; b. IR images after denoising; c. 2-D spectral distribution $|F(u, v)|$; d. horizontal spectral distribution $|F(u, 0)|$; e. vertical spectral distribution $|F(0, v)|$

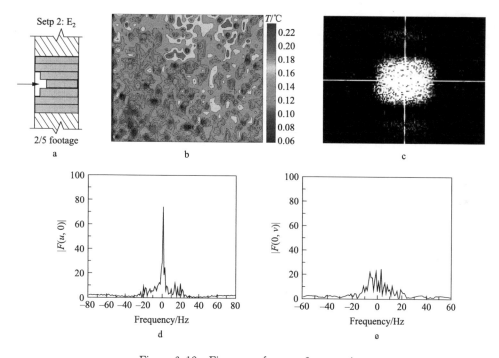

Figure 6.12　Figure set for step 2 excavation

a. excavation diagram; b. IR images after denoising; c. 2-D spectral distribution $|F(u, v)|$; d. horizontal spectral distribution $|F(u, 0)|$; e. vertical spectral distribution $|F(0, v)|$

Figure 6.13a–Figure 6.15a (corresponding to the excavation steps from E_3 to E_5) illustrate the face advancing deep into the model and sliding between the layers took place. In the thermal image E_0 (Figure 6.10b), a random and uniform IRT distribution represents the intact state of the model. In the thermal images E_1 (Figure 6.11b) and E_2 (Figure 6.12b), the IRT distribution was changed to a scattering-random but not uniform pattern, indicating the initiation of the microscopic fractures due to the excavation induced material failure.

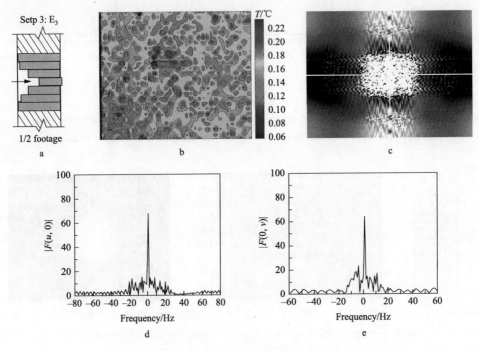

Figure 6.13 Figure set for initial state of the model

a. excavation diagram; b. IR images after denoising; c. 2-D spectral distribution $|F(u, v)|$;
d. horizontal spectral distribution $|F(u, 0)|$; e. vertical spectral distribution $|F(0, v)|$

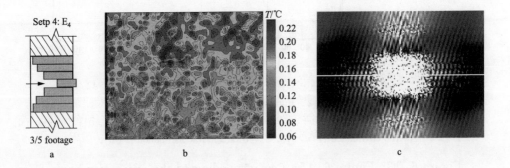

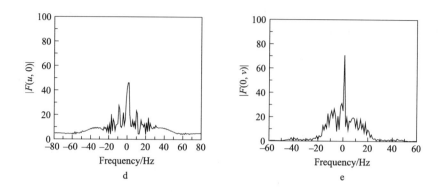

Figure 6.14

a. step 4 excavation; b. IR images after denoising; c. 2-D spectral distribution $|F(u, v)|$;
d. horizontal spectral distribution $|F(u, 0)|$; e. vertical spectral distribution $|F(0, v)|$

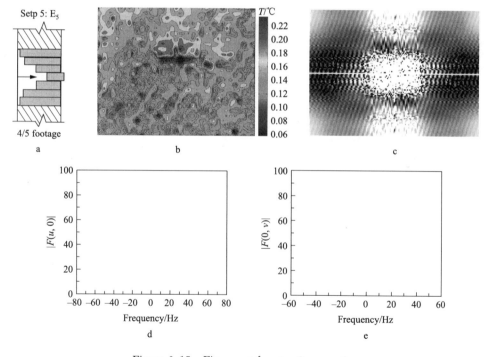

Figure 6.15 Figure set for step 3 excavation

a. excavation diagram; b. IR images after denoising; c. 2-D spectral distribution $|F(u, v)|$;
d. horizontal spectral distribution $|F(u, 0)|$; e. vertical spectral distribution $|F(0, v)|$

In the thermal images E_3 to E_5 (Figure 6.13b-Figure 6.15b), a rectangle-shaped area of low level IRT centered on the roadway excavation zone in deep blue color illustrates the plastic damage caused by the sliding of the rock layers driven by the tunneling impact forces; at the same time, the concentrated IRT distributions above the sliding slabs represent damage localization, and development of these localized IRT distribution depicts the stress path and stress redistribution process by the thermomechanical coupling effects.

Figure 6.16a (E_6) depicts the moment at which the #1 RB was destructed and a small passage was going to cut through (the critical point, with respect to the global maximal value T_{max} on the IRT profile). In the thermal image E_6 (Figure 6.16b), the increased area of the low level IRT indicates the development of the permanent damage by the destruction of the #1 RB and overlying IRT localization indicates the resulting damage. It was noted that the rock layer sliding induced EDZ development and stress redistribution mainly distributed in the overlying strata.

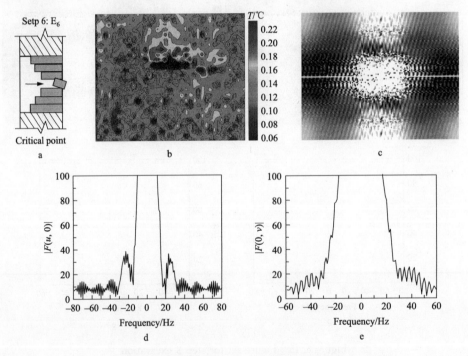

Figure 6.16 Figure set for step 6 excavation
a. excavation diagram; b. IR images after denoising; c. 2-D spectral distribution $|F(u, v)|$; d. horizontal spectral distribution $|F(u, 0)|$; e. vertical spectral distribution $|F(0, v)|$

Because of heterogeneous and plastic nature of the deep soft rocks, the excavation induced stress waves are anisotropic, thus the horizontal and vertical spectra will

Chapter 6　Excavation in horizontal strata · 175 ·

be closely related to the stress path and orientations of the rock strata.

The 2-D and sectional spectra E_0 (Figure 6.12c) can be understood as:

(1) the regular-shaped 2-D spectral distribution centered on the spatial frequency plane, and the randomly distributed sectional spectra with narrow band, indicate the intact state of the model at the beginning of the roadway excavation;

(2) E_1 (Figure 6.13c) and E_2 (Figure 6.14c): the regular-shaped 2-D spectrum on the center with line-formed components across the plane represents microcracking induced stress waves; it was verified by the sectional spectra, from which we see an slightly increased bandwidth and amplitude; it was noticed that the amplitudes for the horizontal spectral (each figure of d) distributions were higher than that of the vertical (each figure of e), indicating the material cracking was more intense in the horizontal than in the vertical during this period of tunneling;

(3) E_3 to E_6 (Figure 6.15c-Figure 6.16c): the 2-D spectrum changed its pattern to the form of a regular-shaped on the center plus dispersed distribution on the periphery, representing localization of the excavation damage, which was further confirmed by the wider band and higher amplitude for the sectional spectra; it should be emphasized that, according to the sectional spectra, one can evaluate the degree and directionality of the excavation damage;

(4) For instance, at POI E_4 (Figure 6.16d,e), the amplitude of the vertical spectrum was higher than that of the horizontal, indicating the material cracking in the vertical direction was stronger than in the horizontal at this time instant. E_6 (Figure 6.16d and e): we see much larger amplitude and the broadest band in both horizontal and vertical directions at the critical point, which was best represented in the frequency domain.

6.4.3　Characterization of the staged excavation

Figures 6.17 to Figure 6.24 show the figure sets with respect to P_0 to P_7 for the staged excavation, which are arranged as the same order as those in Figure 6.10 to Figure 6.16. The understanding and interpretation of these are given as follows.

Diagram P_0 (Figure 6.17a) depicts the time instant when a small passage was perforated in the #1 RB for the first time. P_1 to P_7 (Figure 6.20a and Figure 6.26a) shows the advancement of the staged excavation, i.e., removing one RB at each stage until the roadway excavation was completed in the PFESA model.

In the thermal mage P_0 (Figure 6.19b), a small region of the IRT distribution on the center of the excavation zone represents the small passage cut through for the first time in the #1 RB, and the rest part of the IR imaging surface was deep blue colored (low IRT), indicating the sudden IRT drop due to the unloading impact by the cutting through.

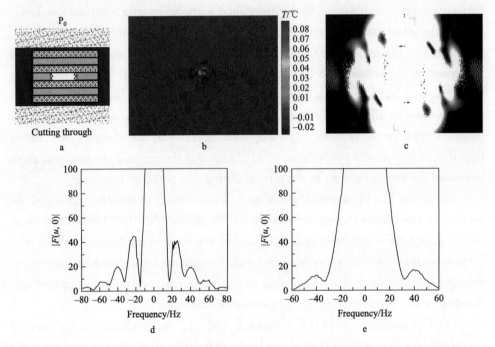

Figure 6.17   Figure set for the excavation stage P_0

a. cutting through on the #1 RB for the first time; b. IR images after denoising; c. 2-D spectral distribution $|F(u, v)|$; d. horizontal spectral distribution $|F(u, 0)|$; e. vertical spectral distribution $|F(0, v)|$

In the thermal images P_1 to P_7 (Figure 6.18b-Figure 6.24b), the expanding area in deep blue color (low IRT) indicates the advancement of the roadway tunneling works, and the adjacent localized IRT distribution represents the EDZ at the specific time instant with respect to P_1 to P_7; the progressive development of the IRT distribution shows the stress paths. It was noticed that from the scale of the IRT localization adjacent to the excavated area (in deep blue color), one can observe and evaluate the advancement of the tunneling and its influence on the neighboring rock mass in terms of the degree and extent of the excavation damage.

For example, at P_1 (complete removal of the #1 RB), the small scale IRT closely nearing the excavated area indicates that the influence of removing #1 RB was limited within the nearby rockmass; at P_2, the IRT distribution extended far away from the excavated area indicates expansion of the influence by removing

Chapter 6 Excavation in horizontal strata · 177 ·

the #2 RB. It was also noticed that the IRT distribution was reducing its scale after P_4, as the excavation approached completion, illustrating the progressive decreasing of the EDZ. From the IR image for P_7, we can see that the IRT distribution was just concentrated on the right side wall in a small size and the IRT distribution elsewhere went back to the pattern similar to the initial state of the excavation E_0, i. e., a random and uniform IRT distribution, indicating the stress redistribution process was getting to an end.

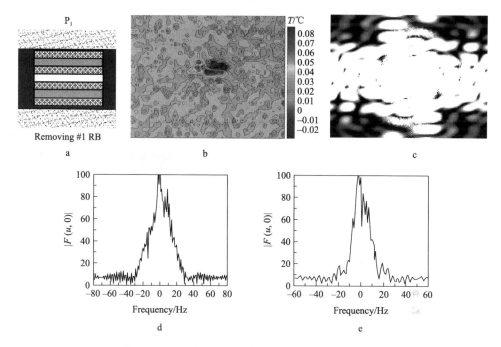

Figure 6.18 Figure set for the excavation stage P_1
a. staged 1 excavation (removing of #1 RB); b. IR images after denoising; c. 2-D spectral distribution $|F(u, v)|$; d. horizontal spectral distribution $|F(u, 0)|$; e. vertical spectral distribution $|F(0, v)|$

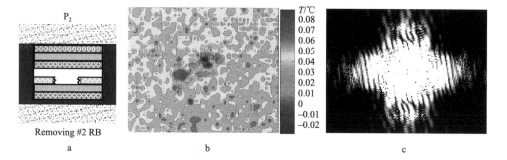

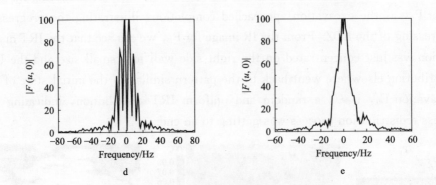

Figure 6.19 Figure set for the excavation stage P_2

a. staged 2 excavation (removing of #2 RB); b. IR images after denoising; c. 2-D spectral distribution $|F(u, v)|$; d. horizontal spectral distribution $|F(u, 0)|$; e. vertical spectral distribution $|F(0, v)|$

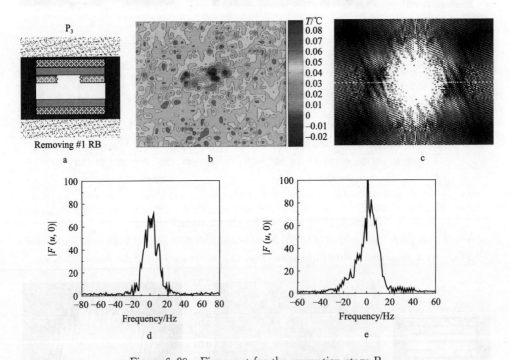

Figure 6.20 Figure set for the excavation stage P_3

a. staged 3 excavation (removing of #3 RB); b. IR images after denoising; c. 2-D spectral distribution $|F(u, v)|$; d. horizontal spectral distribution $|F(u, 0)|$; e. vertical spectral distribution $|F(0, v)|$

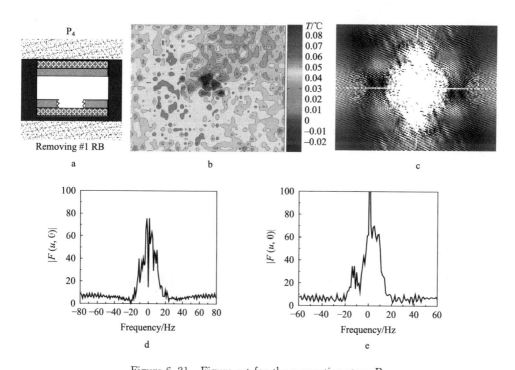

Figure 6.21 Figure set for the excavation stage P_4
a. staged 4 excavation (removing of #4 RB); b. IR images after denoising; c. 2-D spectral distribution $|F(u, v)|$; d. horizontal spectral distribution $|F(u, 0)|$; e. vertical spectral distribution $|F(0, v)|$

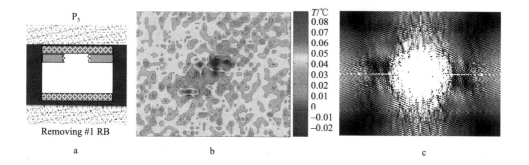

· 180 · Infrared Thermography for Geomechanical Model Test

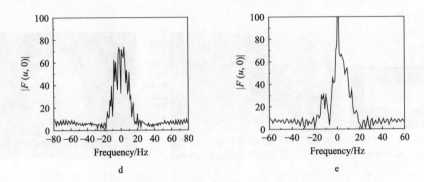

Figure 6.22 Figure set for the excavation stage P_5

a. staged 5 excavation (removing of #5 RB); b. IR images after denoising; c. 2-D spectral distribution $|F(u, v)|$; d. horizontal spectral distribution $|F(u, 0)|$; e. vertical spectral distribution $|F(0, v)|$

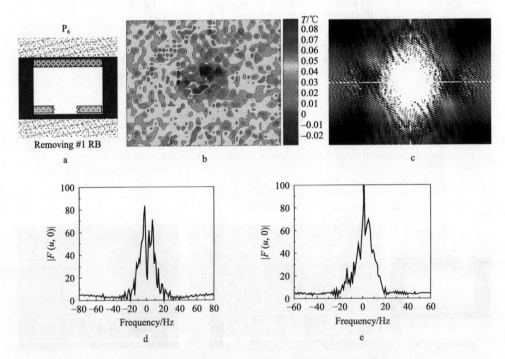

Figure 6.23 Figure set for the excavation stage P_6

a. staged 6 excavation (removing of #6 RB); b. IR images after denoising; c. 2-D spectral distribution $|F(u, v)|$; d. horizontal spectral distribution $|F(u, 0)|$; e. vertical spectral distribution $|F(0, v)|$

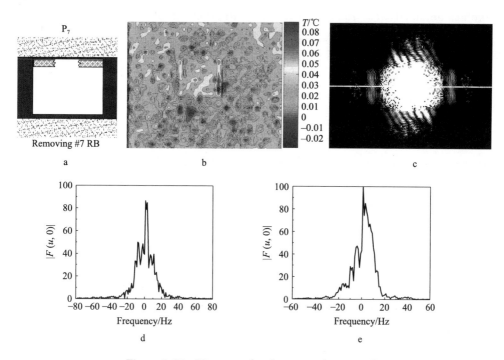

Figure 6.24 Figure set for the excavation stage P_7

a. staged 7 excavation (removing of #7 RB); b. IR images after denoising; c. 2-D spectral distribution $|F(u, v)|$; d. horizontal spectral distribution $|F(u, 0)|$; e. vertical spectral distribution $|F(0, v)|$

There are big differences between the stage 1 and 2 excavations in their 2-D and sectional spectra (see each figure of c, d and e). The characteristics in frequency domain for the phase 2 excavation P_0 to P_7 can be summarized as: ①irregular-shaped 2-D spectrum, in a large scale with high brightness, represents the intense dissipation of the elastic energy induced by cutting through or removal of the RBs, which was described by the sectional spectra with much wider band and higher amplitude. ②the sectional spectra were sensitive to the rock beddings, i.e., after P_2, the vertical spectra had higher amplitude than that of the horizontal, representing that the removal of the RBs resulted in much intensive cracking normal to the rock layers for the horizontally stratified strata.

6.4.4 Summary

Mechanical and structural responses for the horizontally stratified rock mass at great depth under roadway excavation were measured and observed in real time and full field by employing the infrared thermography in our test. The thermographies obtained were further processed with such graphical procedures as data statistics, image segmentation and 2-D DFT, for characterization of the progressive development of the excavation damage in the IR radiation regime, EDZ distribution and spatial frequency domain. Major attainments of this study were concluded as follows.

(1) PFESA method proposed in this study was proved to be an effective means to construct large-scale geological models. With the specimen sized elemental slab as basic unit, it provides a great flexibility in reproduction of the geological and structural features of the stratified rock masses by making these unit in layers at different inclinations.

(2) The mean value of IRT of the infrared images (also referred to as IRT) was used as a measure of the overall rock response to the excavation processes, in terms of the IRT level and its time-marching scheme. The IRT profile increases linearly during the full-face excavation; while fluctuates with multiple local peaks during the staged excavation. At the critical point (destruction of the rock block for the first time, marked by E_6), IRT arrived at the global maximum point. As a result, the full-face excavation and the staged excavation can be treated as the linear and the non-linear processes in the IR regime respectively, from an energy dissipation point of view.

(3) Infrared thermography, incorporated with the noise removal algorithm proposed in this study, have provided a detailed description and characterization of the mechanical and structural responses for the PFESA model under roadway excavation in terms of initiation, propagation and coalescence of damage in a real-time sense and over the entire field especially at an early excavation phase, which otherwise would have been undetectable and unobservable.

(4) The excavation induced stress waves can be characterized by the 2-D Fourier spectrum and its sectional spectral distributions along the horizontal and vertical direction (parallel to and normal to the rock layers respectively), by computing DFT form image matrix of the resultant thermographies. These spectra contribute a great deal in a deepened understanding of the excavation responses by the simplified image features in frequency domain. For instance, the 2-D spectrum characterizes the full-face excavation, the staged excavation and the critical point with distinct spatial distribution patterns; besides, the sectional spectra are sensitive to the rock beddings, and can be used to describe the directional propagation of the stress waves (parallel or normal to the rock strata), by such

indices as the bandwidth or amplitude of the spectral components (related to the scales of fractures and energy dissipation level).

References

Cai M, Kaiser P K, Uno H, et al. 2002. Influence of stress-path on the stress-strain relations of jointed rocks, In: Lin Y, et al (eds). Second International Conference on New Development in Rock Mechanics and Rock Engineering: 60-65.

Chambon P, Corte J F. 1994. Shallow tunnels in cohesionless soil: stability of tunnel face. J. Geotech. Eng., 120(7): 1148-1165.

Connolly M, Copley D. 1990. Thermographic inspection of composite material. Materials Evaluation, 48(12): 1461-1463.

Gonzalez R C, Woods R E, Eddins S L. 2005. Digital Image Processing (English Edition). Beijing: Publishing House of Electronics Industry.

He M C. 2006. Rock mechanics and hazard control in deep mining engineering in China. Rock Mechanics in Underground Construction, ISRM International Symposium 2006 4th Asian Rock Mechanics Symposium: 14-31. Published by World Scientific Publishing Co. Pte. Ltd, 5 Toh Tuck Link, Singapore 596224.

Kaiser P K, Yazici S, Maloney S. 2001. Mining-induced stress change and consequences of stress path on excavation stability-A case study. International Journal of Rock Mechanics and Mining Sciences, 38(2): 167-180.

Kamata G, Masimo H. 2003. Centrifuge model test of tunnel face reinforcement by bolting. Tunneling and Underground Space Technology, 18(2): 205-216.

Kwon S, Lee C S, Cho S J, et al. 2009. An investigation of the damaged zone at the KAERI underground research tunnel. Tunneling and Underground Space Technology, 24(1): 1-13.

Lee Y Z, Schubert W. 2008. Determination of the round length for tunnel excavation in weak rock. Tunneling and Underground Space Technology, 23(3): 221-231.

Li S J, Yu H, Liu Y X, et al. 2008. Results from in-situ monitoring of displacement, bolt load, and disturbed zone of a powerhouse cavern during excavation process. International Journal of Rock Mechanics and Mining Sciences, 45: 1519-1525.

Luong M P. 1990. Infrared thermovision of damage processes in concrete and rock. Engineering Fracture Mechanics, 35(1/2/3): 291-310.

Mark A P. 2003. Introduction to Fourier analysis and Wavelets. Beijing: China Machine Press.

Meguid M A, Sada O, Nunes M A, et al. 2008. Physical modeling of tunnels in soft ground: A review. Tunneling and Underground Space Technology, 23(2): 185-198.

Meola C. 2007. A new approach for estimation of defects detection with infrared thermography. Materials Letters, 61: 747-750.

Nomoto T, Imamura S, Hagiwara T, et al. 1999. Shiel tunnel construction in centrifuge. Journal of Geotechnical and Geoenvironmental Engineering, 125(4), 289-300

Read R S. 2004. 20 years of excavation response studies at AECL's Underground Research Laboratory. International Journal of Rock Mechanics and Mining Sciences, 41(8): 1251-1275.

Ruistuen H, Teufel L W. 1996. Analysis of the influence of stress path on compressibility of weakly cemented sandstones using laboratory experiments and discrete particle models. In: Aubertin M. et al (eds). Proceedings of Second North American Rock Mech. Symposium: 1525-1531

Sharma J S, Bolton M D, Boyle R E. 2001. A new technique for simulation of tunnel excavation in a centrifuge. Geotechnical Testing Journal, 24(4): 343-349.

Steinberger R, Valadas Leitão T I, Ladstätter E, et al. 2006. Infrared thermographic techniques for non-destructive damage characterization of carbon fiber reinforced polymers during tensile fatigue testing. International Journal of Fatigue, 28: 1340-0347.

Tanaka T, Sakai T. 1993. Progressive failure and scale effect of trapdoor problem with granular materials. Soils and Foundations, 33(1): 11-22.

Tang Y G, Kung T C G. 2009. Application of nonlinear optimization technique to back analyses of deep excavation. Computers and Geotechnics, 36: 276-290.

Toubal L, Karama M, Lorrain B. 2006. Damage evolution and infrared thermography in woven composite laminates under fatigue loading. International Journal of Fatigue, 28(12): 1867-1872.

Wu B R, Lee C. 2003. Ground movement and collapse mechanisms induced by tunneling in clayey soil. International Journal of Physical Modeling in Geotechnics, 3(4): 13-27.

Wu L X, Liu S J, Wu Y H, et al. 2006. Precursors of rock fracturing and failure-Part I: IRR image abnormalities. International Journal of Rock Mechanics and Mining Sciences, 43: 473-482.

Wu L X, Wang J Z. 1998. Features of infrared thermal image and radiation temperature of coal rocks loaded. Science in China (Series D), 41(2): 158-164.

Chapter 7　Overloaded tunnel in 45° inclined rocks

7.1　Introduction

Sedimentary rocks cover the majority of earth's surface and are frequently encountered in underground mining. In the sedimentary rocks, two main sources of discontinuities are beddings and joints. The beddings can be assumed continuous over areas greater than that of any designed excavation. The joints however, are typically constrained between beddings. Both beddings and joints are surfaces of relatively low shear strength and negligible tensile strength.

Under the condition of stratified rock masses stretching across the roadway section, the engineering geological behavior during roadway development and operation is mainly controlled by the characteristics of the stratification planes (Fortsakis et al., 2012). Existence of these discontinuities may exert a marked impact on the reduction of rock strength and stiffness of the rocks (Sagong and Bobet, 2002). Stability analysis of the excavations in sedimentary rocks should account for geometrical and mechanical properties of the discontinuities (Tesarsky, 2012).

Extensive researches have been conducted on tunneling, roadway excavation and reinforcement, block caving and stability of the underground caverns in sedimentary rocks including, for example, in-situ tests (Read, 2004; Li et al., 2008); analytical studies (Lydzba et al., 2003); numerical modeling using finite difference method (FEM) (Tsesarsky, 2012), finite element method (FEM) (Golshani et al., 2007; Fortsakis et al., 2012), discrete element method (DEM) (Heuze and Morris, 2007), discontinuous deformation analysis (DDA) (Hatzor and Benary 1998; Tsesarsky and Hatzor 2006; Mazor et al., 2009; Zuo, et al., 2009), and physical model tests (Sharma et al., 2001; Kamata and Masimo, 2003; Liu et al., 2003; Castro et al., 2007; Lee and Schubert, 2008; Shin et al., 2008; Fekete et al., 2010; Zhu, et al., 2011; Li et al., 2013). Well designed experiments and judicious choice of model materials and their response and matching this with stress levels may yield important views into failure modes and mechanisms that are not available from numerical models (Zhu et al., 2011).

In order to capture the geotechnical information about detailed rock conditions and responses, the use of remote sensing equipment is often required during in-situ or laboratory tests. Laser scanning and photogrammetry are two imaging techniques widely used in a tunnel environment. Digital imaging system for determining displacement and strain has been applied in recent decades to a number of geotechnical engineering problems (Lee and Bassett, 2006; Gaich and Potsch, 2008; Birch, 2008), and recently used successfully for monitoring the convergence around cavern in large-scale three-dimensional geomechanical model tests (Zhu et al., 2010, 2011).

The photogrammetry, as reviewed by Fekete et al. (2010), requires supplementary lighting while three-dimensional laser scanning (Lidar) acts as its own source of "illumination" (Kim et al., 2006; Birch, 2008). The Lidar was applied but not limited to the evaluation of rock reinforcement (Gosliga et al., 2006), landslide monitoring (Strouth and Eberhardt, 2005), and stratigraphy modeling (Buckley et al., 2008). Fekete et al. (2010) used improved Lidar in active tunneling environment under dusty, damp, and dark conditions and collected very accurate, high resolution 3-dimensional images of its surroundings.

The advantages of employing non-contact optical vision techniques lie in their ability to represent the structural change by realistic and practical surface models or geometrical features. Usability of the detected geometrical features such as cracks and discontinuities, however, depends on the image resolution and does not has a definite relation to stress redistribution in the surrounding rocks.

Infrared (IR) thermography is another non-contact and remote sensing technique which produces thermal image in real time by detecting electromagnetic waves within infrared wave band (Luong, 1995). Thermal image represents rock response based on the thermal-mechanical coupling effect (Luong, 2007) and does not require supplementary lighting as well.

When processed with proper algorithms, the thermal image will not only be able to detect geometrical features such as crack propagation, but also the static and dynamic friction (He et al., 2010a) which could hardly be observed by the conventional optical visualization techniques. Thermography matrix data set is in fact the IR radiation temperature field on the surface in view induced by energy release of the cracking rocks. The fact that frequency-spectra of the thermal image can represent the seismic wave propagation is the intrinsic advantage of the thermography that the optical imaging technique does not possesses.

In the previous chapters, reported are the geomechanical model test on the roadway excavations in 0°, 45°, 60° and 90° inclined stratified rocks (He et al., 2010a, 2010b; He, 2011; Gong et al., 2013b). However, detecting the large-scale geomechanical model tests on tunnel stability embedded in steeply inclined rocks using infrared thermography was rarely addressed in the published literatures.

Stability of the underground opening increases as the joint set dip is flattened, and decreases when it is steepened (Heuze and Morris, 2007). The lowest strength of the rock masses correspond to the dip angles ranging from 40-50°, evidenced by the systematic studies using large-scale geomechanical model tests and numerical simulations (Zhu and Zhao, 2004).

In this chapter, findings obtained from the experimental investigation on the frictional failure behaviour of the steeply inclined strata with an un-supported tunnel will be presented. The dip of 45° was selected based on the above reviewed findings by Zhu and Zhao (2004). In addition to thermal image analysis with the state-of-the art algorithm, the loading rate and Fourier spectra were utilized to analyze high and low *spatial frequency* precursors on rock failure.

7.2 Experimental

7.2.1 Geomechanical model

The surrounding rocks of the prototype are characterized by complicated engineering geology, developed joints and fissures, and very poor stability. For simulating such structures, a compound reference rock mass model can be used based on the GSI rock mass characterizations system and rock mass structure and discontinuities surface quality (Marinos and Hoek, 2000; Marinos et al., 2005; Fortsakis et al., 2012), as shown in Figure 3.8.

The reference rock mass is composed by the dominant discontinuities (e.g. beddings) and internal rock mass without the persistent discontinuities but contains all the secondary discontinuities. The reference rock mass can be considered as the qualitative sum of the internal rock mass and dominant discontinuity system. The dominant discontinuities are bedding, schistosity and weak surface; the blocky internal rock mass are often interlocked, partially disturbed and undisturbed rock mass consisting of cubical blocks form by the orthogonal intersecting discontinuity sets, namely the secondary discontinuities (Fortsakis et al., 2012).

The field case simulated in this research is a main haulage roadway under operation in QISHAN underground coal mine, located in Xuzhou coal mining district, eastern China. The mining depth of this mine at present is ranging from 300-1000 m and will proceed to greater depth of more than 1000 m in the future. For construction of the large-scale physical model consisting of 45° inclined strata, the field investigation, laboratory specimen tests, dimensional analysis were carried out. Detailed description of the prototype can be found in chapter 3.

Figure 7.1 shows the constructed large-scale geomechanical model for simulating the geological structure in Figure 3.8. The model are composed by total nine strata including one sandstone, four mudstones and four coal seams, having a dimension of 1600 mm×1600 mm×400 mm. These strata are marked by 1-9, and their thickness, the number of the slab layers, and s the geological section are also shown in the figure. An unsupported cubical space with cross section of 250 mm×200 mm and axial length of 400 mm was embedded in the coal seam stratum 4.

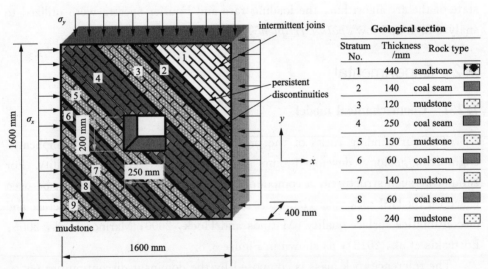

Figure 7.1 Large-scale geomechanical model for simulating an unsupported tunnel embedded in 45° inclined strata subjected to high stresses

7.2.2 Loading path

The two-dimensional loading path was designed to explore the failure mechanisms of the un-supported underground opening under great overburdens and high tectonic stresses. The vertical load was designed to simulate the overburden stress due to the weight of the rock above the cavern. The vertical loads were calculated by the relation, $\sigma_y = \gamma h$, where γ is the generalized unit weight equal to 27 kN/m³ and h (m) is the depth below ground surface. The horizontal load, σ_x, is equal to $\lambda \sigma_y$ where λ is the lateral pressure coefficient, defined by $\lambda = \sigma_x/\sigma_y$, used for characterization of the non-hydrostatic loading.

The applied vertical and horizontal stresses, (σ_y, σ_x), on the boundaries of the physical model were scaled by the force scale factor $\alpha_\sigma = 8$ and the results were reported in Table 7.1. The loading path consists of totally 13 loading stages indexed with capital letters A-N. During the test, the loading stages were applied stepwise equivalent to the overburden depths from 296-2074 m.

Table 7.1 Loading path consisting of totally 13 loading stages

Stress State label	Applied stresses/MPa		Lateral pressure coefficient λ, $\lambda = \sigma_x/\sigma_y$	Actual stresses/MPa		Simulated depth, h /m
	Vertical stress, σ_y	Horizontal stress, σ_x		Vertical stress, σ_y	Horizontal stress, σ_x	
A	1	3.5	3.5	8	28	296
B	1.4	3.6	2.57	11.2	28.8	415
C	1.6	3.8	2.38	12.8	30.4	474
D	2	3.8	1.9	16	30.4	593
E	2.6	4.0	1.54	20.8	32	770
F	3	4.2	1.4	24	33.6	889
G	3.2	4.2	1.31	25.6	33.6	948
H	3.8	4.6	1.21	30.4	36.8	1126
I	4.4	5	1.14	35.2	40	1304
J	5	5.4	1.08	40	43.2	1481
K	5.6	5.8	1.04	44.8	46.4	1659
L	6.2	6	0.97	49.6	48	1837
N	7	6.4	0.91	56	51.2	2074

7.3 Infrared detection

7.3.1 Infrared thermography and imaging procedures

Figure. 7.2 shows the infrared detection of the physical model test. Thermal sequence were acquired by infrared thermography, model TVS-8100 MK Ⅱ, which was cooled during the operation by a built-in refrigerator. The infrared camera works in the passive mode (no extra heat sources used) at wave length of 3.6-4.6 μm, with measuring temperature range of $-40-+300$ ℃; minimum detection temperature difference of 0.025℃; a field of view of 13.6°×18.2°(with standard lens, $f=25$ mm); spatial resolution of 2 mrad; on-line display resolution 240×320 pixels. The raw thermogram were stored in the recorder as digital image of 120×160 pixels for off-line processing.

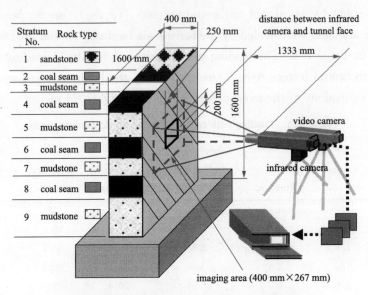

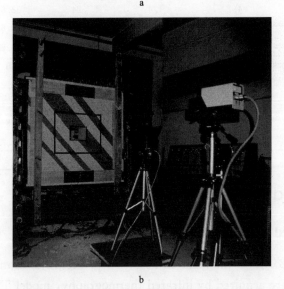

Figure 7.2 Infrared detection of the large-scale geomechanical model test
a. schematic; b. photograph

The infrared camera was fixed to a photographic tripod and placed in the front side of the geological model at a distance of 1333 mm so as to have an imaging area of 400 mm×367 mm indicated by a 400 mm×300 mm red-colored frame of plastic thin tape glued to the model front face. Prior to the test, the camera is initialized and some parameters like emissivity, reflected temperature, air temperature, relative humidity are set up. The emissivity was set to 0.92 for the simulated model rock masses. For 24 hours earlier before the testing, all the instruments were placed in the same room with the physical model, so that the detected infrared temperatures are the temperature variation caused by the external loading. During the test, the image acquisition frequency was set to one frame every four seconds.

7.3.2 Temperature calibration

The methodology on the use of infrared thermography, in general, has two categories: i.e., passive thermography and active thermography. Active thermography uses an external heating device for heating up the testing object under external loading and, therefore, needs to calibrate the obtained temperature increase physically by manned-controlled operation with respect to the 'known heat source'. Passive thermography detects temperature rise of the testing object under external loading without the use of any extra heat sources.

When using active thermography, quantitative analysis of thermal image could be performed by the *physical calibration* of the temperature increment against the applied loads or displacement. In contrast, analysis would be qualitative when using passive thermography. The choice of the modes for using the infrared thermography is related to the geometrical and physical properties of the object in view.

By heating up the object under testing, active thermography can acquire thermal image with large temperature increment. Thus the thermal image has a large dynamic scale and high contrast capable of representing the physical process clearly. Metals and composites generally have a relatively simple constitutive relationship and high thermal conductivity. In this case, the use of infrared thermography in the active mode is preferred.

Active thermography has been widely used but not limited to the cases, for example, damage characterization of the stressed carbon fiber reinforce polymers (Steinberger et al., 2006; Mayr et al., 2011); defect detection of magnetic specimens (Lahiri et al., 2015) and aluminum specimens (Pastor et al., 2008), and quantitative analysis of plastered mosaics (Theodorakeas et al., 2014). However, the active thermography was rarely used for detection of rock materials due to the structural heterogeneity and small thermal conductivity. Thus, when detecting rock materials, the infrared thermography is usually used in a passive mode.

When using passive thermography, although no requirement for the *physical calibration* of the temperature increment, however, finding a reference point mathematically is needed for characterization of the detected temperature variation, which could be referred to '*mathematical calibration*'. The mathematical calibration can be realized by the *image subtraction* algorithm. While the object under detection is subjected to the external loading, the interested features of the infrared sequence are the temperature increment relative to that of the initial state. Taking the first frame of infrared sequence when the object was at the initial state, subtraction of the first frame from the following images obtains the temperature increment relative to the initial state of the object.

The *mathematical calibration* was implemented by using equation which is re-written here, the image subtraction

$$\hat{f}_k(x,y) = f_k(x,y) - f_0(x,y) \qquad (7.1)$$

where, $f_k(x, y)$ represent the image matrix of kth frame in the sequence which is actually the detected temperature field at the kth instant of time; $\hat{f}_k(x,y)$ is the incremental temperature field (also image matrix) at the kth instant obtained by subtraction of the first frame from $f_k(x, y)$; $f_0(x, y)$ is the first frame of the thermal sequence taken at the initial state of the loading; the subscript k is an integer served as frame index; $x=1,2,\cdots, N$ and $y=1,2,\cdots, M$ are the pixel coordinates, and $M=160$ and $N=120$ are the maximum pixel number respectively for the image matrix.

7.3.3 Image processing

The tasks for processing thermal image acquired in the large-scale geomechanical model tests include: ①removal of different types of noises, and ②enhancement of the low-contrast image. The image processing techniques were well established in our previous works (Gong et al., 2013b) and the related algorithms used in this research are summarized in the following.

(1) For removal of the environmental radiation noise, the image subtraction

expressed in Eq. (7.1) was employed. The image subtraction can also be used as the temperature calibration procedure as introduced in section 7.3.2.

(2) For eliminating the salt-and-pepper noise induced by the electronic current in the measurement instruments, median filter was used.

(3) For reduction the additive-periodical noise which may come from the rotating parts in the cooling system embedded in the infrared camera, Gaussian high-pass filters (GHPF) in the frequency domain was utilized.

(4) When detecting a large-scale object with the IR camera working in the passive mode, the raw thermal image will have a small dynamic scope. As a result, the images should be enhanced in order to represent the rock response clearly. In this paper, a morphological enhancement filter, κ_n, proved to be very effective and developed by Gong et al. (2013a), was used.

Detailed discussion on the image process algorithms for the treatment of the low-contrast and noisy thermal images can be found in Gong et al. (2013a, 2013b).

7.4 Fourier analysis

Formation of cracks, emergence of shearing planes and other modes of rapid macroscopic failure in geologic media involve numerous grain scale mechanical interactions generating stress waves (also seismic waves) due to the abrupt release of stored strain energy (Michlmayr et al., 2012). The stress waves in high-frequency band are acoustic emission (AE) or microseismic (MS) signals which carry information about source, including location and mechanism, defined by mode and magnitude (Kao et al., 2011).

The AE and MS events are indicators of rock fracturing or damage as the rock is brought to failure at high stress (Cai and Kaiser, 2005). The abruptly-released energy can also be detected by infrared thermography and the frequency spectra of the thermal sequence can be used to characterize rock responses (He et al., 2010a, 2010b; He, 2011; Gong et al., 2013b). This section presents the principles of performing a quantitative frequency-spectra analysis of thermal image based on Fourier transform.

7.4.1 Stress wave propagation

It is assumed that a wave propagates in the x coordinate direction in a three-dimensional, elastic isotropic continuum. Passage of the wave induces transient displacements $u_x(t)$, $u_y(t)$, $u_z(t)$ at any point in the medium as indicated in Figure 7.3. The essential notion in the concept of a plane wave is that, at any instant in time, displacements at all points in a particular y-z plane are identical, i.e., (u_x, u_y, u_z) are independent of (y, z).

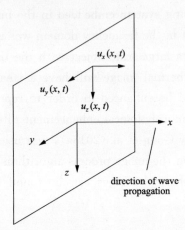

Figure 7.3 Specification of plane waves propagation in the x coordinate direction (after Brady and Brown, 2004)

Alternatively, the definition of a plane wave may be expressed in the form (Brady and Brown, 2004):

$$u_x = u_x(x,t), \ u_y = u_y(x,t), \ u_z = u_z(x,t) \qquad (7.2)$$

The generic form of Eq. (7.2) can be obtained by solving the differential equation of motion for the components of a plane waves

$$u_x = f_1(x - C_P t) + F_1(x + C_P t) \qquad (7.3)$$

$$u_y = f_2(x - C_S t) + F_2(x + C_S t) \text{ and } u_z = f_3(x - C_S t) + F_3(x + C_S t) \qquad (7.4)$$

where,

$$C_P = \{[(1-\nu)/(0.5-\nu)](G/\rho)\}^{\frac{1}{2}} \qquad (7.5)$$

$$C_S = (G/\rho)^{\frac{1}{2}} \qquad (7.6)$$

ν is the Poisson's ratio; G is the shear modulus; ρ is the mass density; the argument $(x - Ct)$ corresponds to a forward progressive wave; the argument $(x + Ct)$ to a backward progressive wave; and C denotes generically the stress wave velocity; The constants C_P and C_S appearing in the Eq. (7.3)-Eq. (7.4) are wave propagation velocities.

Eq. (7.3) describes particle motion which is parallel to the direction of propagation of the wave. Wave propagation at a velocity C_P given by Eq. (7.5) is called *P waves*, or primary or longitudinal waves. Eq. (7.4) describes particle motion which is transverse to the direction of wave propagation. Wave propagation at a velocity C_S given by Eq. (7.6) is called *S waves*, or secondary or shear waves. Natural sources of wave motion normally generate both P and S waves. Transmission of P and S waves in a non-homogeneous medium is subject to internal reflection.

For the case normal incidence on an interface between domains with different elastic properties, an incident P wave generates transmitted and reflected P waves. The ratio of the characteristic impedances, n_P, of the two media for the P waves is defined by (Brady and Brown, 2004)

$$n_P = \rho_2 C_{P2} / \rho_1 C_{P1} \tag{7.7}$$

Similar for the S waves

$$n_S = \rho_2 C_{S2} / \rho_1 C_{S1} \tag{7.8}$$

Note that subscripts 1 and 2 denote different material properties, and the geomechanics convention for the sense of positive stresses and strains is used here.

Suppose the longitudinal stress and particle velocity in the forward incident wave are denoted by magnitude σ_0, V_0, and the corresponding magnitudes in the transmitted and reflected waves are denoted by σ_t, V_t and σ_r, V_r. In the case of one dimensional wave propagation, consider a cylindrical composite bar consists of two segments with different material properties, denoted by subscripts 1 and 2. The reflected stress is,

$$\sigma_r = [(n-1)/(n+1)]\sigma_0 \tag{7.9}$$

The reflected particle velocity is,

$$V_r = -[(n-1)/(n+1)]V_0 \tag{7.10}$$

where, n is a generic form of the ratio of the characteristic impedance of the two media. Eq. (7.9) and Eq. (7.10) can be used to calculate the reflected and transmitted stresses and particle velocities when substitute n with n_P and n_S in these equations. Suppose σ_0 is compressive, for the case $n > 1$, the reflected wave is characterized by a compressive stress; for $n<1$, the reflected wave induces a tensile stress. Thus an important general point to note is that internal reflections of a compressive wave in a medium may give rise to tensile stresses. Details of this problem were discussed by Brady and Brown (2004).

Generation of tensile stresses at free surface and bedding planes provides a plausible mechanism for interbed separation, initial formation of small slip regions, and sliding failure. The dynamic properties of the three prototype rocks were reported in Table 7.2. The ratios of the characteristic impedance n_P and n_S were calculated using Eq. (7.7) and Eq. (7.8). The subscripts in the round brackets for n_P and n_S denote different rock properties, i.e., 1-sandstone, 2-mudstone, and 3-coal seam. It is seen that the characteristic impedance ratios are: $n_{P(2/1)}$ and $n_{S(2/1)} > 1$, $n_{P(3/2)}$ and $n_{S(3/2)} < 1$, and $n_{P(3/1)}$ and $n_{S(3/1)} < 1$. Thus, the reflected waves in the interfaces of sandstone-mudstone, mudstone-coal seam and sandstone-coal seam are characterized by tensile stresses when the incidence is normal and σ_0 is compressive.

Table 7.2 Dynamic properties of the real rocks

	Rock type	Density /(kg/m³)	P-wave velocity /(m/s)	S-wave velocity /(m/s)	Ratio of the characteristic impedance	
					n_P	n_S
1	sandstone	2755	3411.3	1660.3	$n_{P(2/1)} = 1.129$	$n_{S(2/1)} = 1.334$
2	mudstone	2653	4000.0	2300.0	$n_{P(3/2)} = 0.196$	$n_{S(3/2)} = 0.258$
3	coal seam	1429	1456.0	1103.0	$n_{P(3/1)} = 0.221$	$n_{S(3/1)} = 0.345$

According to Eq. (7.7)—Eq. (7.10), at the surface of the opening, $\rho_2 = C_2 = 0$, then both n_P and $n_S = 0$, a compressive pulse is reflected completely as a tensile pulse. It means that violent failure will occur when the propagating waves carry a sufficiently large magnitude of the released energy.

7.4.2 Fourier transform

Stratified rock mass contains dominant and secondary discontinuities, as a result, wave reflection and refraction will be complex. For example, oblique incidence of P and S waves at an interface between dissimilar materials results in more complicated interaction than for normal incidence.

Considering an incident P wave, transmitted and reflected P waves are generated in the usually way. In addition, transmitted and reflected S waves (called PS waves) are produced, i.e., the interface acts as an apparent source for S waves. Similar considerations apply to an incident S wave, which gives rise to SP waves, in addition to the usual transmitted and reflected waves (Kolsky, 1963; Brady and Brown, 2004). As the complexity of the wave motion induced at the interface, characterization of the wave motion in frequency domain could be considered.

The continuous Fourier transform (CFT) of time-domain signal $f(t)$ is defined as (Karris, 2003)

$$F(\omega) = \int f(t) e^{-j\omega t} dt \qquad (7.11)$$

The function $F(\omega)$ is known as the *'Fourier transform'* or the *'Fourier integral'* existing for every value of the *radian frequency* (i. e., *time angular frequency*) ω. The integral $F(\omega)$ is a function of ω ($\omega=2\pi\nu$) or frequency ν (Hz). Eq. (7.11) can also be written in a simple form:

$$F(\omega) = \mathrm{CFT}[f(t)]$$

The Fourier transform is, in general, a complex function. It can be expressed as the sum of its real and imaginary components, or in exponential form, that is, as

$$F(\omega) = \mathrm{Re}\{F(\omega)\} + j\mathrm{Im}\{F(\omega)\} = |F(\omega)| \exp[j\phi(\omega)] \qquad (7.12)$$

where, $|F(\omega)|$ is referred to as *'amplitude spectrum'*; $\phi(\omega) = \arg[F(\omega)]$ is the *'phase spectrum'*.

The Fourier transform of periodic functions produces discrete line spectra with non-zero values only at specific frequencies referred to as *harmonics*. The frequency spectra of a periodic function are multiples of a *fundamental frequency* or *first harmonic* which is the lowest value of the discrete line spectra. However, the frequency spectra for non-periodic functions are continuous. Basically, a non-periodic signal is a continuous function arising from a periodic signal in which the period extends from $-\infty$ to $+\infty$. That is, a non-periodic signal is a function of time with period from $-\infty$ to $+\infty$ (Bracewell, 2000).

7.4.3 Periodicity in time domain

The periodicity of the waves both in time and spatial domains can be expressed by time-and space-related parameters,

$$C = \lambda/T = \lambda\nu \qquad (7.13)$$

where, C is the velocity of the wave; λ is the wavelength, representing the distance that wave travels in a cycle; T is the period of time which is consumed by the wave traveling a wavelength distance and ν is reciprocal of T, referred to *frequency* (or exactly the '*time frequency*') representing the number of times of the wave source vibration in unit time. Considering a harmonic wave propagating along x coordinate direction.

Displacement of a material element in point x at time instant t is given by

$$u(x,t) = A\cos[\omega(t-x/C)] \qquad (7.14)$$

where, ω is the *time angular* frequency, $\omega = 2\pi\nu = 2\pi/T$.

Eq. (7.14) can also be written as

$$u(x,t) = A\cos 2\pi(\nu t - x/\lambda) \qquad (7.15)$$

For a given x value x_0, waveform in Eq. (7.15) is reduced to a function of time variable t only,

$$u(t) = A\cos(2\pi\nu t - 2\pi x_0/\lambda) \qquad (7.16)$$

where, $2\pi x_0/\lambda$ is the initial phase in the time domain.

Eq. (7.16) stands for a material element at point x vibrating at different time as shown in Figure 7.4. In the figure, the cosine curve in solid line denotes the wave form at point x_0, and the dashed line denotes the wave form at point $x = x_0 + \Delta x$. The two wave forms have the same amplitude and frequency, the only difference of which is the initial phase in space. Eq. (7.16) is a periodic function of time with single period. Then its frequency spectrum $U(\omega) = CFT[u(t)]$ has a single line, i.e., the fundamental frequency ω or ν.

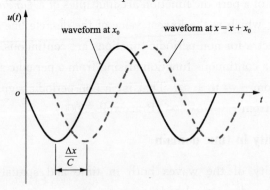

Figure 7.4 Single waveform propagating in time domain, the waveform is expressed by Eq. (7.16)

7.4.4 Periodicity in spatial domain

For a given t value t_0, Eq. (7.15) is reduced to only a function of x coordinates only,

$$u(x) = A\cos(2\pi k x - 2\pi \nu t_0) \tag{7.17}$$

where, $k = 1/\lambda$ is referred to as '*wave number*' or '*spatial frequency*' representing the number of times of the wave source vibration in unit length (Gong et al., 2008); $\breve{\omega} = 2\pi k$ is referred to '*angular frequency*' or '*spatial angular frequency*', and $2\pi \nu t_0$ is the initial phase in space.

Eq. (7.17) stands for a series of material elements with different x coordinates vibrating at a specific time t_0, as shown in Figure 7.5. In the figure, the cosine curve in solid line denotes the wave form at time t_0, and the dashed line denotes the wave form at time $t = t_0 + \Delta t$. The two waves have the same amplitude and spatial frequency, but different initial phase in time. Performing of the Fourier transform on Eq. (7.17) yields '*spatial frequency spectrum*', $U(\breve{\omega})$. Because Eq. (7.17) is also a periodic function of x coordinates with single spatial frequency k. Then $U(\breve{\omega})$ contains only a single component, i.e., the *spatial fundamental frequency* or *spatial harmonic* k.

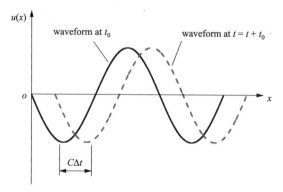

Figure 7.5 Single waveform propagation in space domain, the waveform is expressed by Eq. (7.17)

7.4.5 Physical meaning of the spatial frequency

Eq. (7.16) and Eq. (7.17) represent the same wave with the same amplitude propagating along the t axis and x axis respectively (see Figure 7.4 and Figure 7.5). The temporal frequency in Eq. (7.16), i.e., $\nu = 1/T$, characterizes the periodicity in time domain, whereas the spatial frequency in Eq. (7.17), i.e., $k = 1/\lambda$, characterizes the periodicity in space.

The physical meaning for transform of a waveform signal from time domain to frequency domain has been well understood and the related frequency-spectra analysis is widely used. In the contrast, principles for the Fourier transform of a signal which is a function of space such as Eq. (7.17) and the spatial variable function in the *spatial frequency* domain have yet been less addressed.

Considering the temperature distribution in a thermal image, $f(x,y)$, which is a two-dimensional (2-D) function of the space variables, (x, y), the pixel coordinates. When either of x or y is given, $f(x,y)$ will be reduced into two one-dimensional (1-D) functions, $X(x)$ and $Y(y)$, and their Fourier transform, $X(2\pi k)$, $Y(2\pi k)$ and the related *spatial frequency*, k, will be used in characterization of the seismic waves induced by the fast-rate loading or rock fracture.

The fact that thermography is capable of detecting the seismic wave are based on the following physical principles. The energy carried by the waves propagating through a unit area per unit time is defined as the averaged '*energy flux density*' which is expressed by

$$I = \rho \omega^2 A^2 C/2 \tag{7.18}$$

Part of the wave energy will emit as radiant exitance, M, in the stressed media, which is linked to the temperature field, T, by the Stefan-Boltzmann law,

$$M = \varepsilon \sigma T^4 \tag{7.19}$$

where, ε is the object's emissivity ($0 < \varepsilon < 1$); σ is the Stefan-Boltzmann constant. Propagation of the seismic waves will induce the temperature variation which thus can be captured by the infrared thermography.

7.4.6 Method for spectral analysis

1. Resampling of thermal image

Generally, the 2-D Fourier transform of an image is not suited for the quantitative analysis. Instead, the image matrix $f(x,y)$ could be resampled into 1-D spatial variable function along the pixel coordinates, such as $X(x)$ and $Y(y)$ and then perform 1-D Fourier transform.

This method is proved to be useful for characterization of the frequency-spectra of the thermal image (Gong et al., 2013b). The collected digital thermal image in this research has a matrix of 160×120 pixels. Two representative 1-D spatial series, $x[n]$ ($n=1,2,\cdots; M=160$), and, $y[n]$ ($n=1,2,\cdots; N=120$), were taken along the horizontal and vertical pixel axes across the center of the image plane.

Figure 7.6 shows schematically the sampling scheme. Compute discrete 1-D Fourier transform (DFT) on $x[n]$, the *'horizontal spectrum'* was obtained by 160 points DFT,

$$X(u) = \sum_{n=1}^{M} x[n]\exp(-2\pi unj/M) \tag{7.20}$$

and on $y[n]$, the *'vertical spectrum'* by 120 points DFT,

$$Y(u) = \sum_{n=1}^{N} y[n]\exp(-2\pi vnj/N) \tag{7.21}$$

where, u is the horizontal spatial frequency, Hz; v is the vertical spatial frequency, Hz; reciprocals of u and v equal to the wavelength in the x and y directions respectively.

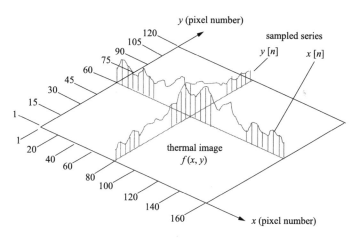

Figure 7.6 Schematic of sampling of the two spatial series x_i ($i=1,2,\cdots, 160$) and y_j ($j=1,2,\cdots, 120$)

2. Characterization of the wave number k

In the 2-D space, $k = \sqrt{u^2+v^2}$ (Hz) represents the spatial frequency for the waves of arbitrary direction, related to the wavelength by $\lambda = 1/k$. The amplitude spectra $|X(u)|$ and $|Y(v)|$ were used to characterize rock responses. It is noted that coordinates of the image matrix $f(x,y)$ are the pixel coordinates. Therefore, the spatial frequency, k (Hz), is the reciprocal of the pixel number coordinates. The resulted wavelength, λ/k, is linked to the real dimension by image resolution. Detailed discussion on this issue could be found in He (2011).

Remember that the unit, Hz, for the spatial frequency, k, u and v, is not related to time and has an extended meaning. The spatial frequency, k, is the reciprocal of wavelength, λ. High spatial frequency corresponds to small wavelength and low spatial frequency to the long wavelength. Failure modes are closely related to the wavelength λ and characteristic length of the tunnel, D. In order to help understand the meaning of the *spatial frequency* $k=1/\lambda$, a critical review of the analytical work on this topic by Yi (1993) was given in the following.

When a plane seismic sine wave encounters a cylindrical underground tunnel with a diameter D, there are generally dynamic stress concentrations and wave reflections at the boundary of the tunnel. Two extreme situations arise depending on the relative size of the tunnel compared to the wave length. One of the extreme situations occurs if the size of the tunnel is negligible as compared to the wave length, that is, $D\lambda^{-1}$ approaches zero; the wave will not be disturbed by the tunnel and wave reflections at the free boundary do not occur. In another words, the loading of the tunnel due to the wave is similar to the static loading.

The other extreme situation occurs if the size of the tunnel is infinite as compared to the wave length, that is, $D\lambda^{-1}$ approaches infinity. In this case, the theory of wave reflection applies and the surrounding rock will experience dynamical impulsive stress. For a situation between the above two extremes where the size of the tunnel D, is comparable to the wave length λ, the concept of wave diffraction is applicable (Yi, 1993). The analytical results are based on the assumption that stress waves propagate in a continuum and isotropic medium. Frequency-spectra analysis based on infrared detection may contribute to understand the situations in complex rock masses.

3. The major frequency

Harmonic frequencies correspond to some major fracture events in the stressed rocks owing to the fact that a point or particle originally at its equilibrium position will be at sinusoidal oscillation when a seismic wave was passing by. Thus the major task for frequency-spectral analysis is to find harmonics existed in the Fourier spectra.

However, identify the harmonics in the spectra of non-periodic function is a tough work because the spectral components, from $-\infty$ to $+\infty$, are fully coupled. Instead, the concepts of predominant frequency, dominant frequency or major frequency are employed by many researchers such as Lu et al. (2012) and Benson et al. (2010). In the following text, we choose the term '*major frequency*' in the spectral analysis.

Figure 7.7 shows schematically three different modes A, B and C for the Fourier spectra. Mode A is the harmonic k_0 only seen in the line spectra for periodic functions; mode B is seen in the continuous spectra for the non-periodic functions; it is a single peak with troughs on the both sides and its peak at k_0 is defined as the major frequency, i. e. , a harmonic (corresponding to a sinusoid) coupled in the continuous spectra.

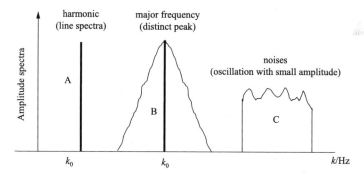

Figure 7.7 Schematic of the typical modes for harmonic and major frequencies

Mode A is the harmonic k_0 only seen in the periodic function; Mode B is the major frequency; Mode C is the noise, and Mode B and C are existed in the continuous spectra for the non-periodic functions

If the spectra contain multiple peaks which are distinct from each other, then they are major frequencies. The lowest major frequency is also referred to as *fundamental frequency* (or first *harmonic*) and the following ones referred to as *second major frequency*, ⋯, and so on. Mode C is the continuous spectra with multiple small oscillations corresponding to the noises. The modes B and C were used in characterization of the frequency-spectra in the following text.

7.5 Loading path and overall rock response

7.5.1 **Energy release index**

As defined in the previous chapters, the mean value of the thermogram matrix, represents the overall energy release rate at a specific instant of time. The statistical mean of the matrix, ⟨IRT⟩, is calculated by,

$$\langle \text{IRT} \rangle = \frac{1}{M} \frac{1}{N} \sum_{x=1}^{M} \sum_{y=1}^{N} f(x,y) \qquad (7.22)$$

Compute the statistical mean on the processed infrared sequence using Eq. (7.22), a time series, $\langle \text{IRT} \rangle_k, k=0,1,2,\cdots$, is generated, where k corresponds to the sampling period of time, i.e., 4 s.

Figure. 7.8 shows the time marching scheme of ⟨IRT⟩ represented by blue line. The time axis is with respect to the thirteen loading stages marked by A-M (see Table 7.2.2). The horizontal and vertical stresses, (σ_x, σ_y), are represented by black and red lines respectively in Figure 7.8 and increased with the stepwise form.

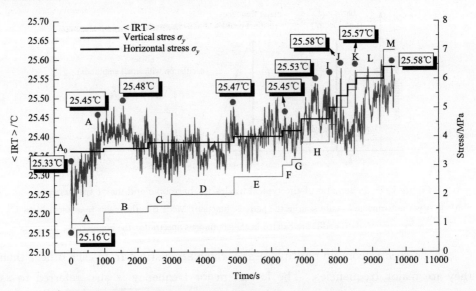

Figure 7.8   Evolution of ⟨IRT⟩ against time with respect to the loading history

During the loading stage A(3.5, 1 MPa) with $\lambda = 3.5$, the ⟨IRT⟩ increases in a quasi-linear manner and after that the ⟨IRT⟩ evolves with a stick-slip oscillation pattern. The loading cases B-G correspond to the overburden depths from 415 to 948 m with larger lateral pressure coefficients, H-M to the depths from 1126-2047 m with smaller later pressure coefficients. Accordingly, ⟨IRT⟩ curve oscillates with small amplitude from B to G and with large amplitude from H to M respectively. It is seen from Figure 7.8 that, in addition to the load level, the stress increment and the period for each loading stage are different, thus the effect of the loading speed on the IRT should be taken into account.

7.5.2 Loading rate

For characterizing the effect of the loading history, three terms, i. e., *incremental stress* $\Delta\sigma$, the *loading period* Δt, and the *loading rate* (or *loading speed*) and were used. As well documented in the elasto-plastic mechanics text books, the definitions for these terms are: ① $\Delta\sigma$ equals to the currently applied stress minus the previously applied stress; ② Δt is the period of time during which the load had been applied and; and ③ the loading rate is defined as $\Delta\sigma/\Delta t$.

According to the loading scheme in Table 7.2, the applied incremental stresses, $\Delta\sigma_x$ and $\Delta\sigma_y$, loading period Δt, horizontal loading rate, $\Delta\sigma_x/\Delta t$, and vertical loading rate, $\Delta\sigma_y/\Delta t$, can be calculated and the results are reported in Table 7.3. Visualization of the data sets in Table 7.3 will be hopeful for understanding the effect of the loading speed on the thermal-energy responses.

Table 7.3 The applied incremental stress, loading period and loading speed

Loading state	$\Delta t/s$	$\Delta\sigma_y/\text{MPa}$	$\Delta\sigma_x/\text{MPa}$	$\Delta\sigma_y/\Delta t/(\text{kPa/s})$	$\Delta\sigma_x/\Delta t/(\text{kPa/s})$
A	980	1	3.5	1.0	3.6
B	1324	0.4	0.1	0.3	0.08
C	692	0.2	0.2	0.29	0.29
D	1890	0.4	0	0.21	0
E	1454	0.6	0.2	0.41	0.14
F	280	0.4	0.2	1.43	0.71
G	288	0.2	0	0.69	0
H	820	0.6	0.4	0.73	0.49
I	224	0.6	0.4	2.68	1.79
J	316	0.6	0.4	1.9	1.27
K	262	0.6	0.4	2.29	1.53
L	808	0.6	0.2	0.74	0.25
M	326	0.8	0.4	2.45	1.23

7.5.3 Characterization of the loading rate effect

Figure. 7.9 illustrates the data sets for the loading rates, $\Delta\sigma_x/\Delta t$ and $\Delta\sigma_y/$

Δt, in Table 7.3. By a careful observation of the histograms in Figure 7.9 and ⟨IRT⟩ curve in Figure 7.9, following insights were obtained:

(1) Loading stage A: corresponding to the overburden depth of 296 m (see Table 7.2),①$\Delta\sigma_x/\Delta t=3.6$ is the fastest loading speed, as a result, the ⟨IRT⟩ curve increased by 0.17℃ as the load was applied; 0.17℃ is a large increment for the ⟨IRT⟩ curve;②since the load level is low, the rock mass response was linear represented by the quasi-linear increase of the ⟨IRT⟩ curve;

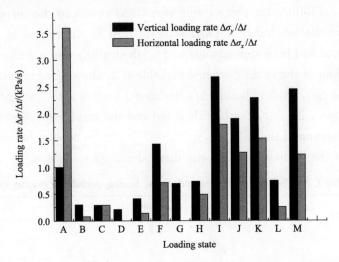

Figure 7.9 Loading speed represented by the horizontal loading rate $\Delta\sigma_x/\Delta t$ and the vertical loading rate $\Delta\sigma_y/\Delta t$

(2) loading stages B-G: corresponding to the overburden depths from 415 to 948 m,①the loading speed was slow compared to those in loading stages I-M;②there would be more time for large portion of the mechanically induced thermal energy being dissipated in the deformation process;③ consequently, fracture propagation would be in a stable manner, and the energy release will be non-violent;④⟨IRT⟩ curve oscillates with small amplitudes.

(3) oading stages H-M: corresponding to the overburden depths from 1126 to 2074 m,①loading speed was faster than those in the previous loading stages in the most cases;②level of the applied overburden stresses are very high;③as a result, there is no enough time for the induced thermal energy to be dissipated, and the large portion of the energy would be stored as elastic energy and released in the fracture events;④fracture propagation would be in an unstable manner, and the energy release will be violent;⑤⟨IRT⟩ oscillates with large amplitudes.

Evolution of the ⟨IRT⟩ curve against time reveals the stick-slip nature of the steeply inclined stratified rocks. Periods for the ⟨IRT⟩ curve oscillation decrease with the increase of the loading speed. Peaks in the ⟨IRT⟩ curve represent the following phenomena:

(1) the stick phase corresponding to the critical static friction during which a lot of heat was produced;

(2) the energy carried by the propagating stress waves produced by ① the suddenly applied load at the beginning of each of the loading steps, and ② impacts by the major interlayer sliding events;

(3) significant energy release during the macroscopic failure event;

(4) dynamic impact created by the fast loading speed with high-level overburdens.

On the contrary, troughs in ⟨IRT⟩ curve represent the sudden drops of the IRT caused by the following phenomena:

(1) loosening of the rocks under the action of the interlayer slip;

(2) bulking deformation of the rock due to opening of the existing cracks;

(3) macroscopic failure due to the crack coalescence;

(4) tensile failure events;

(5) dynamic sliding under the condition of the mobilized cohesion and small friction strength.

7.6 Results and discussions

7.6.1 Terms and approach

Method introduced in section 7.4.3 was used to analyze the Fourier spectra of the denoised thermal sequence. In order to characterize the frequency-spectra of the thermal sequence, the frequency axis were divided into three bands, i.e., "*low band*" within frequency range of 1-2 Hz; "*high band*" within 2-3 Hz, and "*ultra-high band*" within the frequency range larger than 3 Hz. In the following text, nine typical thermal images along with their amplitude spectra will be analyzed in accordance with the loading cases A (the first loading stage), B-E-F-H (with small loading rate) and I-J-K-M (with fast loading rate).

7.6.2 Spectra characterization of loading state A

Figure 7.10 and Figure 7.11 show the thermal images (the figure in upper panel) and their Fourier spectra (two figures in the lower panel) for the loading stage A(3.5, 1 MPa) and $\lambda=3.5$. The applied horizontal stress is much higher than that of the vertical stress creating a stress condition where the tectonic stress is dominant and the surrounding rock masses of the opening would undergo a highly unbalanced stress state.

1. Loading case A_0

Point A_0 in the $\langle$IRT$\rangle$ curve (see Figure 7.8) indicates the instant when the loads were firstly applied on the model, the corresponding thermal image and Fourier spectra are given in Figure 7.10. In the thermal image A_0, high and low IRT distribute with a scattering-random pattern, indicating the fact that:

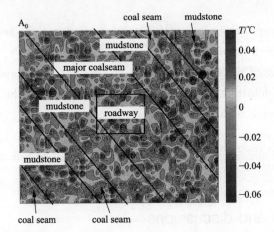

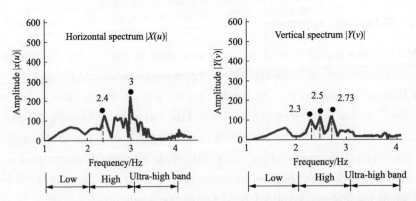

Figure 7.10 Typical thermal images and Fourier spectra in loading stage A(3.5, 1 MPa) and $\lambda=3.5$; loading speed $\Delta\sigma_x/\Delta t=1$ and $\Delta\sigma_y/\Delta t=3.6$ (the largest vertical loading speed)

A_0-thermal image and the spectra for A_0 correspond to the instant when the loads were firstly applied

(1) rock response is elastic under the low-level loads, and

(2) there was no damage to the rock masses at this moment.

For comprehension of the anisotropic behavior of different rock mass, the roadway section and rock strata are marked by black lines in the image corresponding to the imaging area in the physical model (see Figure 7.2). The stratum 4 containing the opening is referred to as *major coal seam* in the following context.

In the horizontal spectrum $|X(u)|$ A_0 (see Figure 7.10), there are two high-band major components at 2.4 and 3 Hz with minor amplitude. In the vertical spectrum A_0, there are three high-band major components with minor amplitude at 2.3, 2.5 and 2.73 Hz, representing three stress waves propagating in the horizontal direction. The Fourier spectra for A_0 indicate the fact that:

(1) the Fourier spectra are more sensitive to the external load and the stress variation can be detected readily by the major components;

(2) the amplitudes for these major components are small, at the moment A_0 there was no noticeable damage to the rock masses as shown in the thermal image A_0.

Although the loading stage A has the fastest vertical loading speed, i.e., $\Delta\sigma_y/\Delta t = 3.6$, it just excited some small-amplitude harmonic waves and the effect of the small-amplitude seismic waves on the rock response is negligible.

2. Loading case A_1

A_1 (marked by a red point) is a representative point in the late phase of the first loading stage as indicated in the ⟨IRT⟩ curve (see Figure 7.8). The thermal image and its Fourier spectra are given in Figure 7.11. It is the third peak in ⟨IRT⟩ curve, representing a continued accumulation of the elastic energy of the rock under loading. It is seen in the image A_1 that the belt-like high IRT distribution is more prominent in the major coal seam indicating the stress concentration in the coal seam and stronger static friction along the interfaces between the coal and mudstone.

The IRT distribution, however, is still in a scattering and random manner, representing the elastic rock response. In the horizontal spectrum A_1, '*frequency shift*' could be observed, for example, a major frequency 1.38 Hz occurred in the low band. However, the amplitude for the low-band major frequency is small, indicating an event with large influence extent in the horizontal direction.

The vertical spectrum for A_1 has two low-band components at 1.39 and 1.98 Hz, and three high-band components at 2.33, 2.6 and 2.73 Hz. The increase of the numbers of the major components represents more harmonic waves propagating across the imaging plane in the vertical direction. It is noted that the horizontal and vertical spectrum has five major components respectively, representing the isotropic rock responses in the loading stage A.

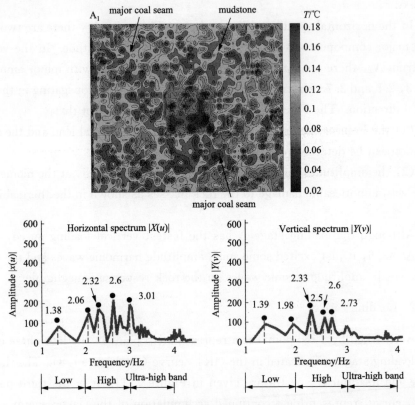

Figure 7.11 Typical thermal images and Fourier spectra in loading stage A(3.5, 1 MPa) and $\lambda=3.5$; loading speed $\Delta\sigma_x/\Delta t =1$ and $\Delta\sigma_y/\Delta t =3.6$ (the largest vertical loading speed)

A_1-thermal image and the spectra for point A_1 in the loading stage A

3. Frequency shift

The term '*frequency shift*' phenomenon refers specifically to the phenomenon that some of the major frequencies are moved from high band to the low band as a significant material fracture event imminent. Read et al. (1995) reported the research on frequency-spectra analysis of acoustic emission (AE) signal monitored in laboratory triaxial experiments on Darley Dale sandstone samples.

The findings showed that the major frequencies were in the high band with low amplitude prior to peak stress; however, after the peak stress, the amplitude increased by an order of magnitude and the frequencies were moved to lower band. The frequency shift as the precursory warnings was also found in many researches, for instance, in monitoring the in-situ rock bursts (Brady and Leighton, 1997; Lu et al., 2012), roof falls (Shen et al., 2008), earthquake induced ground motion (Chernov and Sokolov, 1999), and laboratory simulated rock bursts (Philip et al., 2010; Lu et al., 2012).

As discussed in section 7.6.4, the frequency used in this research is actually the *spatial frequency* which is inversely proportional to the wavelength. Two extremes, i. e. , $D\lambda^{-1} = 0$ and $D\lambda^{-1} = \infty$, are correspond to the static loading and dynamic loading applied to the tunnel. In order to have a clear graphical representation of a wave, a sinusoidal S wave was used to demonstrate the meaning of the parameter of $D\lambda^{-1}$ used in the condition of this research, shown in Figure 7.12.

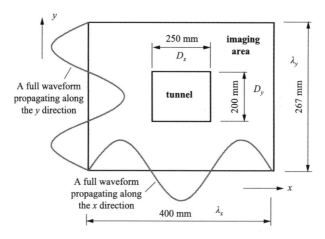

Figure 7.12 Schematic of the maximum wavelengths of the stress waves propagating in x- and y-coordinates directions; the two waves used for the illustration are sinusoidal S waves

The dimensions for the tunnel and imaging area (see Figure 7.2) was schematically drawn in Figure 7.12 in which a full waveform propagating in the x coordinate direction (*x-waveform*) and a full waveform propagating in the y coordinate direction (*y-waveform*) are shown. The maximum wavelength of the x-waveform, λ_x, equals to 400 mm and the maximum wavelength of the y-waveform, λ_y, equals to 267 mm corresponding to the two side lengths of the imaging area, which are the maximum wavelengths that the infrared thermography could capture.

As shown in Figure 7.12, the parameter $D\lambda^{-1}$ for the x-waveform, λ_x is

$$D_x \lambda_x^{-1} = 250/400 = 0.63 \tag{7.23}$$

and for the y-waveform, λ_y is

$$D_y \lambda_y^{-1} = 200/267 = 0.75 \tag{7.24}$$

where, $D_x = 400$ mm and $D_y = 267$ mm are the horizontal and vertical lengths of the imaging area. Eq. (7.23) and Eq. (7.24) give the smallest values of the parameter $D\lambda^{-1}$, far away from the extreme, $D\lambda^{-1} = 0$. Thus the stress waves with low spatial frequency k (or long wavelength) represent the tunnel-wide dynamic events.

For example, in Figure 7.10 and Figure 7.11, thermal image A_0 has a scattering-random IRT distribution indicating the elastic-isotropic behaviour and there are no low-band frequencies its Fourier spectra. In the contrast, thermal image A_1 has a belt-like IRT distribution and the IRT distribution borders, i.e., the static friction surfaces, were at a tunnel-wide scale. The phenomenon of the tunnel-wide static friction was precisely captured by the Fourier transform of the thermal image by the low frequency components, i.e., the *frequency shift* phenomena.

It is worth mentioning that according to Eq. (7.18), the amplitude spectra, $|X(u)|$ and $|Y(v)|$, are related to the IRT variation. Small amplitude represents low energy release. It is seen from Figure 7.11 that all the major frequencies in the spectra A_1 have small amplitude, representing the dynamic event did not cause a substantial damage to the rock masses.

7.6.3 Characterization of loading cases with slow loading rate

1. Loading case B

Figure 7.13 to Figure 7.16 show the figure sets for the typical loading cases, B, E, F and H, with slow loading rates. In the loading case B(3.6, 1.4 MPa) with $\lambda=2.57$ being the second largest lateral pressure coefficient, both the horizontal and vertical rate were small, i.e., $\Delta\sigma_y/\Delta t = 0.3$ and $\Delta\sigma_x/\Delta t = 0.08$.

The representative point with red dot for loading case B in the ⟨IRT⟩ curve (see Figure. 7.8) has a marked IRT increased by 25.48°. The thermal image B

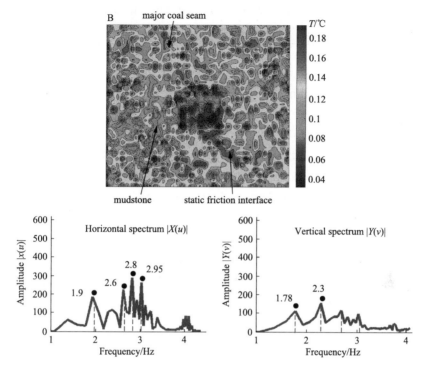

Figure 7.13 Thermal images and Fourier spectra corresponding to the typical loading stages with slow loading rates
figure set B(3.6, 1.4 MPa), $\lambda=2.57$, $\Delta\sigma_y/\Delta t=0.3$ and $\Delta\sigma_x/\Delta t=0.08$

and its Fourier spectra are shown in Figure. 7.13. High IRT distribution in the upper-right part of the plane and an oblique high IRT thin strip represent the intense static interlayer friction. Low IRT region in the left part of the plane indicates the relaxation after the localized slip along the bedding planes.

In the horizontal spectrum B (lower panel in Figure 7.13), we can see that:

(1) One low-band component at 1.9 Hz with small amplitude represents the tunnel-wide event did not cause a substantial damage due to the small energy release level;

(2) Three high-band components at 2.6, 2.8 and 2.95 Hz with moderate amplitude, representing the fact that the influence by the dynamic loading induced by the seismic wave propagation was moderate, i. e., the static friction event shown in the thermal image B.

In the vertical spectrum B, there are one low-band component at 1.78 Hz and one high-band component at 2.3 Hz, with small amplitude respectively, representing the fact that their effect were negligible in the vertical direction.

Because the loading rates is relatively small in the loading case C ($\Delta\sigma_y/\Delta t = \Delta\sigma_y/\Delta t = 0.29$) and the loading case D ($\Delta\sigma_y/\Delta t = 0.21$ and $\Delta\sigma_y/\Delta t = 0$), we omit the related discussions due to the fact that loading with small incremental stresses and low loading speed did not cause a marked changes in rock behaviour.

2. Loading case E

In the loading case E, applied stresses: (4.0, 2.6 MPa) with $\lambda = 1.54$, equivalent to the real overburden depths of 770 m approaching to the critical depth in deep mining; the vertical loading speed, $\Delta\sigma_y/\Delta t$, equals to 0.41 faster than those in the previous loading cases; the incremental vertical load, $\Delta\sigma_y$, equals to 0.6 MPa larger than the former (see Table 7.3).

It is seen in Figure 7.8 that the ⟨IRT⟩ curve peaks immediately at the beginning of the loading stage E, marked by a red dot. The thermal image and its Fourier spectra corresponding to this point were selected for the discussion. In the thermal image E (in Figure. 7.14), the major coal seam was stressed heavily represented by high IRT and the clear borders between the strata indicating the intense static friction. The localized large-scale IRT zones on the two-side walls and floor represent the plastic damage. On the lower part of the plane, low IRT zone represents the local interlayer slip.

The unconfined compressive strength (UCS) and Young's modulus for the model coal rock are 3 MPa and 0.61 GPa respectively (see Table 7.1) while for the model mudstone are 5 MPa and 2.62 MPa respectively. The model coal, therefore, is much softer than the model mudstone. Hence, the coal seam is easy to deform and will experience stronger stress concentration and static friction. In the horizontal spectrum E, in addition to the three high-band components, one ultra-high-band major frequency at 3.25 Hz indicates the dynamic loading to the tunnel as discussed in section 7.4.3. The ultra-high-band component was the precursor predicting an imminent fracture event.

In the loading case F, applied stresses: (4.2, 3 MPa) with $\lambda = 1.4$ equivalent to the overburden depth of 889 m belonging to deep mining; incremental vertical load $\Delta\sigma_y = 0.6$ MPa. Both the vertical and horizontal loading speed, i.e., $\Delta\sigma_y/\Delta t = 1.43$ and $\Delta\sigma_x/\Delta t = 0.71$, were faster than those in the previous loading cases. The incremental vertical load, $\Delta\sigma_y = 0.6$ MPa, was larger as well. It is seen from the thermal image F (in Figure 7.15) that the IRT distribution is

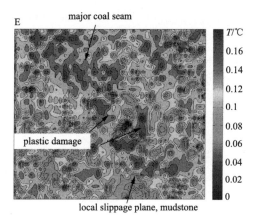

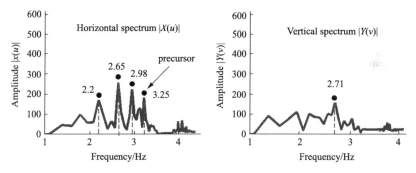

Figure 7.14 Thermal images and Fourier spectra corresponding to the typical loading stages with slow loading rates

figure set E(4.0, 2.6 MPa), $\lambda=1.54$, $\Delta\sigma_y/\Delta t=0.41$ and $\Delta\sigma_x/\Delta t=0.14$

divided into distinct high and low IRT belts parallel to the bedding planes.

It is very interesting that the IRT has an inversed distribution, i.e., the coal seam strata are low IRT belts while mudstone strata are high IRT belts. In comparison to the previous loading cases (such as E), the coal seam strata are high IRT belts while the mudstone strata are low IRT belts. The abnormal IRT distribution in the thermal image F reveals the fact that:

(1) Before the loading stage F, the hard rock strata exerted forces on the soft rock strata, i.e., the coal strata were highly stressed and undergoes the intense static friction both within the rock mass and along the bedding planes;

(2) At the loading stage F, the overstressed coal strata deform, collapse and slid along the interlayer surfaces exerting inversely the load on the hard rock strata.

In the horizontal spectrum F (in Figure 7.15), the abnormal spectral precursors include: two low-band components, one at 1.38 Hz with highest amplitude,

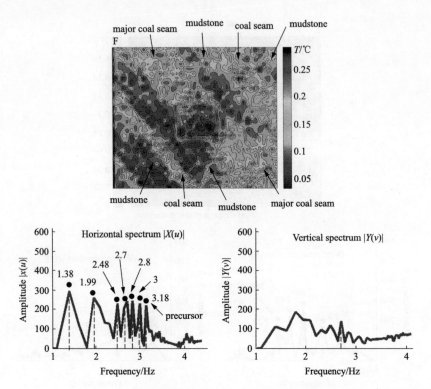

Figure 7.15 Thermal images and Fourier spectra corresponding to the typical loading stages with slow loading rates

figure set F(4.2, 3 MPa), $\lambda=1.4$, $\Delta\sigma_y/\Delta t=1.43$ and $\Delta\sigma_x/\Delta t=0.71$

another at 1.99 Hz with the second highest amplitude; and one ultra-high-bane component at 3.18 Hz with moderate amplitude. The ultra-high-bane component is the precursor indicating the fact that the fracture induced loading is dynamic. The two low-band components with large amplitude represent the phenomenon of *frequency shift*, indicating the tunnel-wide fracture event as shown in the thermal image F, i. e. , all the interlayer slip represented by the IRT belts across the plane. In the vertical spectrum F, the spectrum oscillates with small amplitude in the stochastic manner, according to section 7.4.3, no major frequencies could be identified in the vertical spectrum.

In the loading stage H, applied stresses: (4.6, 3.8 MPa) with $\lambda=1.21$ and $\Delta\sigma_y=0.6$ MPa, equivalent to the overburden depth of 1126 m belonging to mining at great depth. Despite both the vertical and horizontal loading speed are slow, i. e. , $\Delta\sigma_y/\Delta t=0.73$ and $\Delta\sigma_x/\Delta t=0.49$, the large overburden stress and incremental stress could also induce a significant damage on the surrounding rocks.

In the thermal image H (upper panel in Figure 7.16), low IRT distribution over most of the plane represents the relaxation effect of the rock due to the interlayer sliding. High IRT belt in the upper-left plane represents the intense internal rock friction in the major coal seam. In the horizontal spectrum H, there are two load-band major components with very high amplitude at 1.3 and 1.93 Hz and one high-band component at 3 Hz, indicating three sinusoidal stress waves propagating along the horizontal direction. The frequency shift phenomenon in the horizontal spectrum indicates a significant energy release due to the frictional sliding at tunnel wide scale. The vertical spectrum H belongs to the continuous spectrum in which no major component exists.

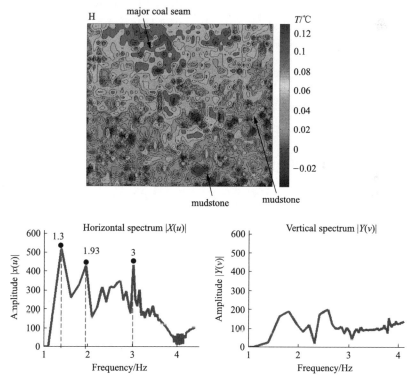

Figure 7.16　Thermal images and Fourier spectra corresponding to the typical loading stages with slow loading rates

figure set H(4.6, 3.8 MPa), $\lambda=1.21$, $\Delta\sigma_y/\Delta t=0.73$ and $\Delta\sigma_x/\Delta t=0.49$

7.6.4 Characterization of loading cases with fast loading rate

1. Loading case I

Four loading stages, I, J, K and M, were selected to characterize the thermal behaviour under the condition of high loading rate and great overburden depth. In the loading stage I, applied stresses: (5, 4.4 MPa) with $\lambda=1.14$, equivalent to the overburden depth of 1304 m. Both the vertical and horizontal loading speed are relatively fast, i.e., $\Delta\sigma_y/\Delta t = 2.68$ and $\Delta\sigma_x/\Delta t = 1.79$. The incremental vertical load, $\Delta\sigma_y = 0.6$ MPa, is also large among others.

A dramatic increase of IRT was induced by the applied loads in the loading stage I, indicated by the red dot in the ⟨IRT⟩ curve (Figure 7.8). In the thermal image I (upper panel in Figure 7.17), like the IRT distribution in the image F (see Figure 7.15), the coal strata were low IRT belts while mudstone were high

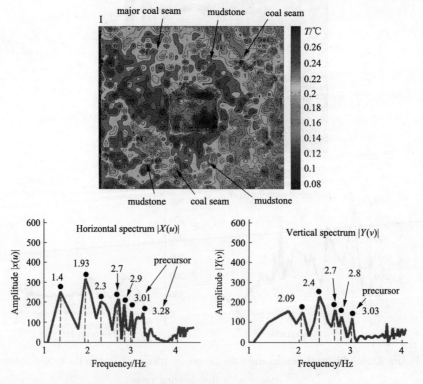

Figure 7.17 Thermal images and Fourier spectra corresponding to the typical loading stages with fast loading rates

figure set I(5, 5.4 MPa), $\lambda=1.14$, $\Delta\sigma_y/\Delta t = 2.68$ and $\Delta\sigma_x/\Delta t = 1.79$

IRT belts, representing the fact that the mudstone strata were stressed heavily due to the sliding of the fractured coal seams. Note that for the steeply inclined rock, deformation mechanism of different zones around periphery of an opening may be very different.

Thermal image I (upper panel in Figure 7.17) illustrates the partition deformation mechanisms. For instance, in the major coal seam containing the tunnel, the upper portion undergoes large deformation and intense internal friction represented by high IRT, whereas, the lower portion is stable evidenced by low IRT distribution.

In the horizontal spectrum, two low-band components at 1.4 and 1.93 Hz represent the macroscopic scale interlayer slip event; two ultra-high band components at 3.01 and 3.28 Hz indicate the event was dynamic in horizontal direction. It was noted that the vertical spectrum I also has an ultra-high band component at 3.03 Hz indicating the fact that the fracture development in the vertical direction was unstable and violent.

2. Loading case J

In the loading case J, applied stresses: (5.4, 5 MPa) with $\lambda=1.08$ approaching to the hydrostatic loading state, $\Delta\sigma_y=0.6$ MPa, and the vertical load equivalent to the overburden depth of 1481 m. The vertical and horizontal loading speed are still fast, i.e., $\Delta\sigma_y/\Delta t=1.9$ and $\Delta\sigma_x/\Delta t=1.27$.

A dramatic increase of the IRT was induced during the loading, indicated by the red dot in the ⟨IRT⟩ curve (see Figure 7.8). In the thermal image J (upper panel in Figure 7.18), the upper-half plane is the high IRT region representing the compressive deformation of the roof under the great overburden. The localized high IRT distribution on the two side walls and floor represents the convergence of the tunnel section and floor heave. In the horizontal spectrum J, the ultra-high band component at 3.18 Hz indicates the fact that rock failure in horizontal direction was dynamic and violent.

3. Loading case K

In the loading stage K, applied stresses: (5.8, 5.6 MPa) with $\lambda=1.04$, almost the hydrostatic loading state, the incremental vertical load, $\Delta\sigma_y=0.6$ MPa and the vertical load equivalent to the overburden depth of 1459 m. The vertical and horizontal loading speed were fast, i.e., $\Delta\sigma_y/\Delta t=2.29$ and $\Delta\sigma_x/\Delta t=$

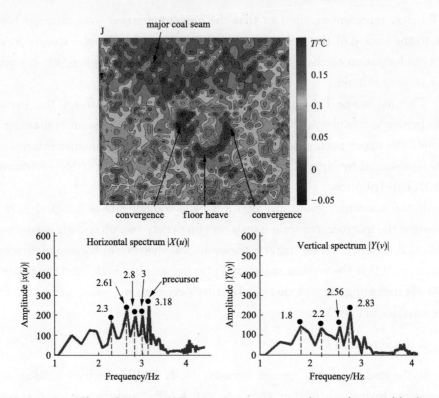

Figure 7.18　Thermal images and Fourier spectra corresponding to the typical loading stages with fast loading rates

figure set J(5.4, 5 MPa), $\lambda=1.08$, $\Delta\sigma_y/\Delta t =1.9$ and $\Delta\sigma_x/\Delta t =1.27$

1.53.

Dramatic increase of the IRT was induced as the loads were applied, indicated by the red dot in the ⟨IRT⟩ curve (see Figure 7.18). In the thermal image K (in Figure 7.19), the coal strata are low IRT regions and mudstone strata are high IRT regions, showing the similar failure mechanism in the loading cases I and F.

A partition failure mechanism was seen clearly in the major coal seam containing the opening. The upper half of the major seam was over stressed while the lower half was stable with regions localized around the opening being over loaded. In the horizontal spectrum K, the low-band components at 1.4 and 1.8 Hz represent the tunnel-wide interlayer slip, and the ultra-high band component at 3.23 Hz represents the dynamic nature of the slip.

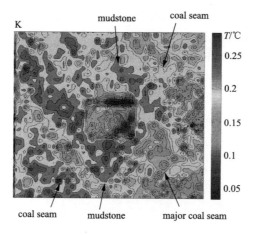
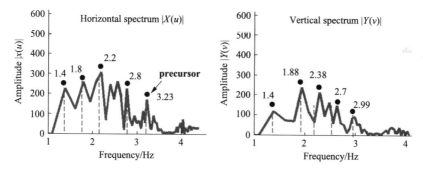

Figure 7.19 Thermal images and Fourier spectra corresponding to the typical loading stages with fast loading rates

figure set K(5.8, 5.6 MPa), $\lambda=1.04$, $\Delta\sigma_y/\Delta t=2.29$ and $\Delta\sigma_x/\Delta t=1.53$

4. Loading case M

In the loading case M, applied stresses: (6.4, 7 MPa) with $\lambda=0.91$, equivalent to the overburden depth of 2074 m. It is the first time the vertical load is larger than the horizontal load. The vertical and horizontal loading speed are fast, i. e., $\Delta\sigma_y/\Delta t=2.45$ and $\Delta\sigma_x/\Delta t=1.23$, with the largest incremental vertical load, $\Delta\sigma_y=0.8$ MPa. Under the great overburden and large incremental vertical stress and fast loading rate, a significant energy release was represented by the peaks (marked by a red dot) in the ⟨IRT⟩ curve (see Figure 7.8).

In the thermal image M (upper panel in Figure 7.20), like the loading cases I, K and F, the coal seams are low-IRT regions while mudstone strata are high-IRT regions. The distinct borders separating the different IRT regions indicate the sliding surfaces. In the horizontal spectrum M, the low-band component at 1.36 Hz with ultra-high amplitude demonstrates the tunnel-wide scale of the interlayer slip. The ultra-high band component at 3.43 Hz indicates the sliding event is violent and dynamic in horizontal direction. It is noted that in the vertical spectrum M, the low-band component at 1.8 Hz with high amplitude and ultra-high component at 3.1 Hz demonstrate the event is tunnel-wide and dynamic in vertical direction.

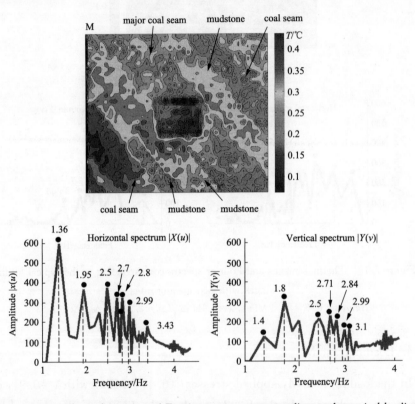

Figure 7.20 Thermal images and Fourier spectra corresponding to the typical loading stages with fast loading rates

figure set M(6.4, 7 MPa), $\lambda=0.91$, $\Delta\sigma_y/\Delta t = 2.45$ and $\Delta\sigma_x/\Delta t = 1.23$

7.6.5 Discussions

The loading path in the test contains 13 cases A-M. These loading cases can be classified into two categories, i.e., the first group contains the loading cases A-G with the overburden depths less than 1000 m, and the second group contains the cases H-M with the overburden depths deeper than 1000 m. Most of the cases in the first group has a relatively small vertical loading speed, i. e., $\Delta\sigma_y/\Delta t \leqslant 1$, whereas, the second group has a relatively larger vertical loading speed, i. e., $\Delta\sigma_y/\Delta t > 2$.

Over two different loading groups, the 45° inclined strata with an embedded tunnel exhibit distinct behaviour which was captured by the thermal images and their Fourier spectra. Although the detailed analyses were conducted in section 7.5 and 7.6, it might take time to capture the key points from the long text. In order to help understand the critical findings of this research, a brief discussion may be needed.

Thermal image illustrates the structural behavior by such image features as colors, borders and edges that separate different zones in the thermal image. By calibration of the temperature variation mathematically (see section 7.3.2), temperature scale of the separated zones represents temperature variation of the stressed rocks relative to the initial state. Pseudo-colors are used to highlight the temperature scales.

Hot (or positive) colors stand for high IRT distribution and cool (or negative) colors for the low IRT distribution. A great number of studies demonstrated that positive IRT corresponds to high stress level due to friction, shearing or stress concentration; negative IRT corresponds to low stress level caused by tensile cracking, dynamic sliding with low frictional strength, opening of the existing cracks, stress release, and unloading, etc., (Wu et al., 2006; Luong, 2007; Pastor et al., 2008; Gong et al., 2013a, 2013b; Grinzato et al., 2004).

The spatial Fourier spectra, $|X(u)|$ and $|Y(u)|$, and spatial frequency, $k = (u^2 + v^2)^{1/2}$ (Hz) along with the wavelength $\lambda = 1/k$ (m) are employed to characterize the frequency-spectra features of the thermal image. The wavelength λ or spatial frequency k has a considerable influence on the failure modes of a tunnel with a characteristic dimension, D. As discussed in section 7.4.3 and section 7.6, high spatial frequency represents that the loading or fracture event is dynamic.

Thus, high spatial frequency can be used as a precursor for the imminent unstable and violent failure. Low frequency component with high amplitude, i. e., '*frequency shift*', in the time series, for example, the AE signal, represents the imminent macroscopic failure such as the rock bursts (Benson et al., 2010; Lu et al., 2012). Whereas, in this research, the low spatial frequency component with high amplitude represents the fact that influencing extent of the fracture was at a tunnel-wide scale (see Figure 7.9). The infrared thermography can well capture the stress wave propagations induced by the fast loading speed or rock fracture, evidenced by the Fourier spectra analysis in this research.

Typical thermal images and their Fourier spectra, selected from the loading cases A_0-A_1 (see Figure 7.7 and Figure 7.8), B-E-F-H (see Figure 7.10 to Figure 7.13), and I-J-K-M (see Figure 7.14 to Figure 7.17), were analyzed in section 7.6, and the improved understanding of the thermal behaviour for the steeply inclined rocks embedded with an un-supported opening could be briefed in what follows.

(1) Two thermal images and their Fourier spectra A_0 and A_1 selected from the loading case A. The scattering-random IRT distribution depicts the elastic behaviour of the rock masses; major components or harmonics in the Fourier spectra are small in amplitude indicating the fact that the dynamic effect can be negligible, despite the loading stage A has the largest horizontal loading speed (see Figure 7.9).

(2) In the thermal images, for loading cases of B-E-F-H with small loading rates and low overburden depth, the coal strata were over stressed, represented by the high IRT, while the mudstone strata were less stressed represented by low IRT, such as the cases B and E (see Figure 7.10 to Figure 7.13). For loading cases I-J-K-M with fast loading rates and great overburden depth, the IRT distribution is inversed, i.e., the mudstone strata were over stressed indicated by high IRT while the coal strata were less stressed indicated by low IRT (see Figure 7.14 to Figure 7.17). The inversed IRT distribution of the rock strata with different physical properties could be taken as an abnormal IRT *precursor* indicating the dynamic interlayer slip under the conditions of small frictional strength or the loss of the cohesion caused by the collapse of the overstressed coal seams with smaller strength as compared to the mudstone.

(3) The low-frequency major components can be found in almost all the loading cases despite the loading speed or overburden depth. However, in the loading cases with fast loading rate such as the case F, or great overburden depth such as the case H, the low-frequency components usually have large amplitude. Therefore, the low-frequency components may represent, ①low-frequency components with small amplitude may indicate the sphere of influence that the stress redistribution extends. For example, in the thermal images for the loading case E (see Figure 7.11), the high IRT belt across the plane represents the static inter-

layer friction which is indicated by the low-band component at 1.9 Hz with small amplitude; ②low-frequency components with high amplitude may indicate a tunnel-wide dynamic sliding event. For example, the thermal image for the loading stage F (see Figure 7.12), as analyzed in section 7.6.2, represents the dynamic interlayer sliding.

(4) The high-frequency major components can be found only in the loading cases that the rock at the critical deformation state or undergoing dynamic sliding event. For instance, in the loading case E (see Figure 7.11), the interlayer static friction between the coal seam and mudstone was at the critical state, the ultra-high frequency component at 3.25 Hz indicates the imminent sliding event in the loading case F (see Figure 7.12). In the subsequent loading case E, ultra-high frequency component at 3.18 Hz indicates the dynamic nature of the interlayer friction.

7.6.6 Summary

Large-scale geomechanical model test was conducted to investigate stability of an un-supported tunnel with rectangular cross section embedded in 45° inclined alternating strata of sandstone, mudstone and coal seam. The stress path consists of 13 loading cases which are divided into two groups: the loading cases A-G with overburden depths from 296-948 m and small loading speed (defined as $\Delta\sigma/\Delta t$), and the loading cases H-M with overburden depths from 1126-2047 m and fast loading speed. Infrared thermography, incorporated with the advanced image processing and Fourier transform, was employed to characterize the rock responses.

The averaged infrared radiation temperature field, $\langle IRT \rangle$, represents the overall energy release rate for the stressed rock masses. The $\langle IRT \rangle$ curve has a short linear phase during the first loading stage, and after that oscillates with different periods and amplitude, illustrating the stick-slip behaviour of the steeply inclined rock masses under stressing. Overburden depth and loading speed have a significant impact on the evolution pattern of the $\langle IRT \rangle$ curve. Under small overburden depth and low loading speed, the $\langle IRT \rangle$ curve oscillates with longer period and small amplitude; whereas, under great overburden depth and fast loading speed, with shorter period and large amplitude.

The processed thermal image best represents rock behaviour by two major IRT distribution patterns. For loading cases with small loading rates and low overburden depth, the coal strata were over stressed indicated by high IRT distribution while the mudstone strata were less stressed represented by low IRT. For loading cases with fast loading rate and great overburden depth, the IRT distribution is inversed, i. e. , the mudstone strata were over stressed indicated by high IRT while the coal strata were less stressed indicated by low IRT. The first IRT pattern indicates the static interlayer friction and the second one indicates the dynamic interlayer friction.

The spatial Fourier spectra, $|X(u)|$ and $|Y(u)|$, and spatial frequency, k (Hz), are employed to characterize the frequency-spectra features of the thermal image. The ultra-high spatial frequency component with large amplitude is a precursor for predicting the imminent dynamic interlayer slip or fracture, and for indicating the dynamic nature of an on-going event. The low spatial frequency component may be served as ①a precursor of a tunnel-wide scale event such as the interlayer slip if the amplitude was high, and ②an indicator of the tunnel-wide sphere of influence that the stress redistribution extends if the amplitude was small or moderate.

References

Brady B T, Leighton F W. 1997. Seismicity anomaly prior to a moderate rock burst, a case study. Int. J. Rock Mech. Min. Sci & Geomech. Abstr. , 14, 127-132.

Bracewell R N. 2000. The Fourier Transform and its Applications. Third Edition. New York: McGraw-Hill Higher Education.

Brady B H G, Brown E T. 2004. Rock mechanics for underground mining. New York: Kluwer Academic Publishers.

Birch J S. 2008. Using 3DM analyst mine mapping suite for underground mapping. In: Tonon F (ed). Proceedings of Laser and Photogrammetric Methods for Rock Tunnel Characterization Workshop. 42nd US Rock Mechanics Symposium ARMA, San Francisco, June 28-29, Cdrom.

Buckley S, Howell J, Enge H, et al. 2008. Terrestrial laser scanning in geology: data acquisition, processing and accuracy considerations. Journal of the Geological Society, London, 165(3), 625-638.

Benson P M, Vinciguerra S, Meredith P G, et al. 2010. Spatio-temperal evolution of volcano seismicity: A laboratory study. Earth Planetary Sci. Lett, 297, 315-323.

Chernov Y K, Sokolov V Y. 1999. Correlation of seismic intensity with Fourier acceleration spectra. Physics and Chemistry of the Earth (A), 24, 523-528.

Cai M, Kaiser P K. 2005. Assessment of excavation damaged zone using a micromechanics model. Tunnelling and Underground Space Technology, 20, 301-310.

Castro R, Trueman R, Halim A. 2007. A study of isolated draw zones in block caving mines by means of a large 3D physical model. International Journal of Rock Mechanics and Sciences, 44(6), 860-70.

Fekete S, Diederichs M, Lato M. 2010. Geotechnical and operational applications for 3-dimensional laser scanning in drill and blast tunnels. Tunnelling and Underground Space Technology, 25, 614-628.

Fortsakis P, Nikas K, Marinos V, et al. 2012. Anisotropic behavior of stratified rock masses in tunneling. Eng. Geology, 141-142, 74-83.

Geng N G, Yu P, Deng M D, et al. 1998. The simulated experimental studies on cause of thermal infrared precursor of earthquake. Earthquake, 18, 83-88.

Grinzato E, Marinetti S, Bison P G, et al. 2004. Comparison of ultrasonic velocity and IR thermography for the characterization of stones. Infrared Phys. & Technol, 46, 63-68.

Gosliga F V, Lindenbergn R, Pfeifer N. 2006. Deformation analysis of a bored tunnel by means of terrestrial laser scanning. In: Image Engineering and Vision Metrology. ISPRS Commission, 36: 167-172.

Golshani A, Oda M, Okui Y, et al. 2007. Numerical simulation of the excavation damaged zone around an opening in brittle rock. International Journal of Rock Mechanics and Sciences, 44, 835-845.

Gaich A, Potsch M. 2008. Computer vision for rock mass characterization in underground excavations. In: Tonon F (ed). Proc. Laser and Photogrammetric Methods for Rock Tunnel Characterization. 42nd US Rock Mechanics Symposium ARMA, San Francisco, June 28-29.

Gong W L, Zhao H Y, An L Q, et al. 2008. Temporal and spatial analysis of infrared images from water jet in frequency domain based on DFT. Journal of Beijing University of Aeronautics and Astronautics, 34(6), 690-694.

Gong W L, Gong Y X, Long A F. 2013a. Multi-filter analysis of infrared images from the excavation experiment in horizontally stratified rock. Infrar. Phys. Technol., 56, 57-68.

Gong W L, Wang J, Gong Y X, et al. 2013b. Thermography analysis of a roadway excavation experiment in 60° inclined stratified rocks. International Journal of Rock Mechanics and Sciences, 60, 134-147.

He M C, Gong W L, Zhai H M, et al. 2010a. Physical modeling of deep ground excavation in geologically horizontally strata based on infrared thermography. Tunnelling and Underground Space Technology, 25, 366-376.

He M C, Jia X N, Gong W L, et al. 2010b. Physical modeling of an underground roadway excavation vertically stratified rock using infrared thermography. International Journal of Rock Mechanics and Sciences, 47, 1212-1221.

He M C. 2011. Physical modeling of an underground roadway excavation in geologically 45° inclined rock using infrared thermography. Eng. Geol, 121(3-4), 165-176.

Hatzor Y H, Benary R. 1998. The stability of a laminated Voussoir beam: Back analysis of a historic roof collapse using DDA. International Journal of Rock Mechanics and Sciences, 2(12), 165-181.

Heuze F E, Morris J P. 2007. Insights into ground shock in jointed rocks and the response of structures there-in. International Journal of Rock Mechanics and Sciences, 44, 647-676.

Kolsky H. 1963. Stress Waves in Solids. New York: Dover.

Kamata G, Masimo H. 2003. Centrifuge model test of tunnel face reinforcement by bolting. Tunnelling and Underground Space Technology, 18(2), 205.

Karris S T. 2003. Signals and systems with MATLAB applications, second ed. Fremont, California: Orchard Publications.

Kim C, Ghanma M, Habib A. 2006. Integration of photogrammetric and LIDAR data for realistic 3-dimensional model generation. In: 1st International Workshop on Mobile Geospatial Augmented Reality, Banff, Alberta, Ganada, May 29-30. Cdrom.

Kao C S, Carvalho F C S, Labuz J F. 2011. Micromechanisms of fracture from acoustic emission. International Journal of Rock Mechanics and Sciences, 48, 666-673.

Lahiri B B, Bagavathiappan S, Sebastian L T, et al. 2015. Effect of non-magnetic in magnetic specimens on defect detection sensitivity using active infrared thermography. Infrar. Phys. Technol., 68, 52-60.

Lee Y J, Bassett R H. 2006. Application of a photogrammetric technique to a model tunnel. Tunnelling and Underground Space Technology, 21(1), 79-65.

Li S J, Yu H, Liu Y X, et al. 2008. Results from in situ monitoring of displacement, bolt load, and disturbed zone of a power house cavern during excavation process. International Journal of Rock Mechanics and Sciences, 45, 1519-1525.

Li S C, Hu C, Li L P, et al. 2013. Bidirectional construction process mechanics for tunnels in dipping layered formation. Tunnelling and Underground Space Technology, 36, 57-65.

Liu J, Feng X T, Ding X L, et al. 2003. Stability assessment of the Three-gorges Dam foundation, China using physical and numerical modeling-part I. Physical model tests. International Journal of Rock Mechanics and Sciences, 40(5), 609-31.

Lee Y Z, Schubert W. 2008. Determination of the length for tunnel excavation in weak rock. Tunnelling and Underground Space Technology, 23, 221-231.

Lu C P, Dou L M, Liu B, et al. 2012. Microseismic low-frequency precursor effect of bursting failure of coal and rock. J. Applied Geophysics, 79, 55-63.

Long M P. 1995. Infrared thermographic scanning of fatigue in metals. Nuclear Engineering and Design, 158, 363-376.

Lydzba D, Pietruszczak S, Shao J F. 2003. On anisotropy of stratified rocks, homogenization and fabric tensor approach. Computers and Geotechnics, 30, 289-302.

Luong M P. 2007. Introducing infrared thermography in soil dynamics. Infrared Phys. Technol., 49, 306-311.

Marinos P, Hoek E. 2000. GSI, a geologically friendly tool for rock mass strength estimation. Proceedings of the GeoEng2000 at the International Conference on Geotechnical and Geological Engineering, Melbourne, Australia: 1422-1466.

Marinos V, Marinos P, Hoek E. 2005. The geological strength index, applications and limitations. Bulletin of Engineering Geology and the Environment, 64, 55-56.

Mazor D B, Hatzor Y H, Dershowitz W S. 2009. Modeling mechanical layering effects on stability of underground openings in jointed sedimentary rocks. International Journal of Rock Mechanics and Sciences, 46, 262-271.

Mayr G, Plank B, Sekelja J, et al. 2011. Active thermography as a quantitative method for non-destructive evaluation of porous carbon fiber reinforced polymers. NDT & E Int., 44(7), 537-543.

Michlmayr G, Cohen D, Or D. 2012. Sources and characteristics of acoustic emissions from mechanically stress geologic granular media-A review. Earth Sci. Reviews, 112, 97-114.

Pastor M L, Balandraud X, Grédiac M, et al. 2008. Applying infrared thermography to study the heating of 2024-T3 aluminum specimens under fatigue loading. Infrared Phys. Technol., 51, 505-515.

Philip M B, Sergio V, Philip G M, et al., 2010. Spatio-temporal evolution of volcano seismicity, a laboratory study. Earth and Planetary Science Letters, 297, 315-323.

Read M D, Ayling M R, Meredith P G, et al. 1995. Microcracking during triaxial deformation of porous rocks monitored by changes in rock physical properties, II. Pore volumometry and acoustic emission measurements on water saturated rocks. Tectonophysics, 245 (3-4), 223-235.

Read R S. 2004. 20 years of excavation response studies at AECL's Underground Research Laboratory. International Journal of Rock Mechanics and Sciences, 41, 1251-1275.

Sagong M, Bobet A. 2002. Coalescence of multiple flaws in a rock-model material in uniaxial compression. International Journal of Rock Mechanics and Sciences, 39, 229-241.

Sharma J S, Bolton M D, Boyle R E. 2001. A new technique for simulation of tunnel excavation in a centrifuge. Geotech. Test. J., 24(4), 343-349.

Shen B, King A, Guo H. 2008. Displacement, stress and seismicity in roadway roofs during mining-induced failure. International Journal of Rock Mechanics and Sciences, 45, 682-688.

Strouth A, Eberhardt F. 2005. The use of LiDER to overcome rock slope hazard data collection challenges at Afternoon Creek, Washington. In: 41st US Symposium on Rock Mechanics, Golden, Colordo. American Rock Mechanics Association, Cdrom.

Steinberger R, Leitão T I V, Ladstätter E, et al. 2006. Infrared thermographic techniques for non-destructive damage characterization of carbon fiber reinforced polymers during tensile fatigue testing. Int. J. Fatig., 28, 1340-1347.

Shin J H, Choi Y K, Kwon O Y, et al. 2008. Model testing for pipe-reinforced tunnel heading in a granular soil. Tunnelling and Underground Space Technology, 23(3), 241-50.

Tesarsky M. 2012. Deformation mechanisms and stability analysis of unmined sedimentary rocks in the shallow subsurface. Eng. Geol. , 133-134, 16-29.

Tsesarsky M, Hatzor Y H. 2006. Tunnel roof deflection in blocky rock masses as a function of joint spacing and friction-A parametric study using discontinuous deformation analysis (DDA). Tunnelling and Underground Space Technology, 21(1), 29-45.

Tsesarsky M. 2012. Deformation mechanisms and stability analysis of unmined sedimentary rocks in the shallow subsurface. Eng. Geology, 133-134, 16-29.

Theodorakeas P, Avdelidis N P, Cheilakou E, et al. 2014. Quantitative analysis of plastered mosaics by means of active infrared thermography. Construct. Build Mater, 73: 417-425.

Wu L X, Liu S J, Wu Y H, et al. 2006. Precursors for rock fracturing and failure-Part Ⅰ, IRR image abnormalities. International Journal of Rock Mechanics and Sciences, 43, 473-482.

Yi X P. 1993. Dynamic response and design of support elements in rock burst conditions. Kingston, Ontario, Canada: Queen's University Ph. D. thesis.

Zhu W S, Zhao J. 2004. Stability Analysis and Modeling of Underground Excavations in Fractured Rocks. New York: Elsevier Ltd.

Zuo J P, Peng S P, Li Y J, et al. 2009. Investigation of karst collapse based on 3-D seismic technique and DDA method at Xieqiao coal mine. Chin. Int. J. Coal Geol, 78, 276-287.

Zhu W S, Zhang Q B, Zhu H H, et al. 2010. Large-scale geomechanical model testing of an underground cavern group in a true three-dimensional (3-D) stress state. Can. Geotech. J., 47, 935-946.

Zhu W S, Li Y, Li S C, et al. 2011. Quasi-three-dimensional physical model tests on a cavern complex under high in-situ stresses. International Journal of Rock Mechanics and Sciences, 48, 199-209.

Chapter 8　Overloaded tunnel in horizontal strata

8.1　Introduction

With fast growing demands for natural resources, more and more deep-seated tunnels and openings are being planned and constructed (Sagong and Bobet, 2002). The appraisal of tunnel convergence in the case of stratified rock masses is much more complex since the dominant discontinuities may lead to highly anisotropic rock responses (Fortsakis et al., 2012). For understanding the nonlinear behavior controlled by bedding planes, numerous experimental investigations have been carried out on deformation and failure process of the stratified rock masses with an embedded excavation. However, accuracy and reliability of the obtained information rely heavily on the data acquisition method employed in the experiment and data processing technique after the experiment.

As for data acquisition approaches, the contact monitoring technique, such as strain gauges and displacement meters, have been successfully applied both in situ and at laboratory scale, but testing results are easy to be affected by development of rock mass damage in the field experiment. Acoustic emission (AE) is an effective means for the non-intrusive detection which can detect the spatial distribution of rock damage by the sources of microseismic events, such as visualization of the EDZ (excavation damaged zone) in the mine-by experiment (MBE) (Eberhardt et al., 1997,1998; Martin et al., 1997; Maxwell et al., 1998; Read et al., 1998; Read, 2004). However, due to its point-monitoring mechanism, it spatial resolution is much lower as compared with the imaging testing approaches.

Apart from the non-intrusive testing and remote sensing, major advantages for the infrared thermography include: ①representing the intrinsic dissipation of the energy due to the thermomechanical coupling; ②detecting the object configuration without the requirement of supplementary lignting; and ③ capable of detecting the static and dynamic friction which would hardly be detected by the else imaging approaches. Low signal to noise ratio and low contrast are the major limitations for analysis and interpretation of thermal images (Suganthi and Ramakrishnan, 2014).

Quality of the infrared images is even worse when detecting a large-scale object with the infrared camera working in a passive mode, i. e. , without extra heat sources used (Gong et al., 2013a). Image denoising and enhancement is, therefore, required for a proper analysis and interpretation of an IR image.

In recent decades, infrared thermography are employed for detecting roadway excavation and rock failure process in the large-scale sedimentary rock model experiments with the stratifications inclined at angles of 0° (He et al., 2009, 2010), 45° (He, 2011), 60° (Gong et al., 2013b) and 90° (He et al., 2010).

This chapter will introduce the development of an algorithm for processing the noisy and low-contrast thermograms obtained from the geomechanical model test reported in reference (He et al., 2009). Rock mass response analysis using infrared sequence processed using the new algorithm and comparisons to the previous processing of the same sequence in the work (He et al., 2009) and to photographs were carried out, to validate the effectiveness of the new algorithm and improve our understanding about nonlinear behavior of the simulated roadway tunnel in the horizontally stratified rock masses.

8.2 Experimental

8.2.1 Geological model

The prototype of this research is also a main haulage roadway under operation in Qishan underground coal mine, located in Xuzhou mining district, in Jiangsu province, eastern China, with mining depth from 300-1000 m will proceed to greater depth of more than 1000 m in the near future. The geological formation of the mine is composed by horizontal coal bed between successive layers of sandstone and mudstone. Documentation of the dimensional analysis and similarity materials are omitted here as it is already presented in chapter 3-7.

Figure 8.1 shows the constructed geological model having a dimension of 1600 mm×1600 mm×400 mm. The model consists of total nine horizontal strata including one sandstone, four mudstones and four coal seams. These rock strata are marked by 1-9, and their thickness, number of the slab layers and geological section are also listed in Figure 8.1. A cubical space with a dimension of 250 mm wide and 200 mm high and 400 mm long was pre-excavated in the model. The dominant discontinuities are the interface between the strata of different rock types. Detailed description of the model rock fabrication could be found in reference (He et al., 2009).

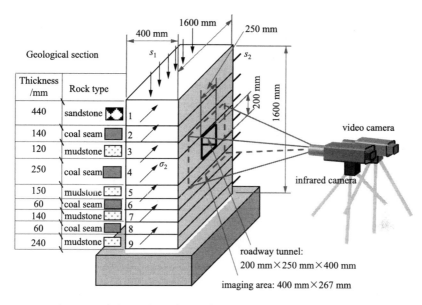

Figure 8.1 Schematic of the geological model, thermal imaging, and video camera recording

8.2.2 Loading scheme

During the test, vertical stress σ_1 and horizontal stress σ_2 were uniformly applied on the top and two side boundaries of the model see (Figure 8.1) by the testing machine. Figure 8.2 shows the stress path which is composed by two phases: hydrostatic loading and non-hydrostatic loading.

Over the hydrostatic loading, the vertical and lateral stresses were increased slowly with the same magnitude from zero to 0.8 MPa, and then held constant for 25 minutes; after that the two stresses continued to increase to 1.4 MPa and then held constant for a while to make the bedding planes closed. During the non-hydrostatic loading, the vertical stress was increased to 5 MPa while the lateral stress held at 1.4 MPa. Then the vertical load was increased to the maximum value of 6 MPa while the lateral pressure was increased to 2 MPa and then to 4 MPa finally.

A table beside the stress path in Figure 8.2 reports the lateral pressure coefficient λ in accordance to the load levels (indicated with round brackets beside the

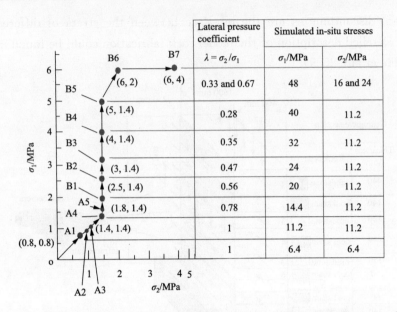

Figure 8.2 Loading scheme and lateral coefficient $\lambda=\sigma_2/\sigma_1$, force scale factor equals to 8

solid red points). The simulated in-situ stresses, equal to the product of the applied stresses and the force scale factor [equal to 8 (He et al., 2009)]. The lateral pressure coefficient is defined as $\lambda=\sigma_2/\sigma_1$ and smaller λ indicates more unbalanced stress states.

8.2.3 Infrared detection

1. Infrared thermography

IR images are acquired by an infrared thermography, model TVS-8100 MK II which was cooled and works in a passive mode (no extra heat sources used) at wave length of 3.6-4.6 μm, with measuring temperature range of -40 to $+300$℃; minimum detection temperature difference of 0.025℃; a field of view of 13.6°×18.2°/25 mm; spatial resolution of 2 mrad; on-line display resolution 240×320 pixels. The raw thermogram were stored in the computer as digital image of 120×160 pixels for off-line processing.

2. Testing procedure

The IR camera was fixed to a photographic tripod and placed in the front side of the geological model at a distance of 1333 mm so as to have an imaging area of

400 mm×367 mm indicated by a 400 mm×300 mm red-colored frame of plastic tape glued to the model front face as shown in Figure 8.3. Prior to the test, the camera is initialized and some parameters like emissivity, reflected temperature, air temperature, relative humidity are set up. The emissivity was set to 0.92 for the simulated model rock masses as used in our tests (He et al., 2009, 2010a, 2010b; He, 2011; Gong et al., 2013). During the test, a video camera with the resolution of 1024×1024 pixels was also placed in the front side of the geological model to record deformation process of the simulated roadway tunnel as shown in Figure 8.1 and Figure 8.3.

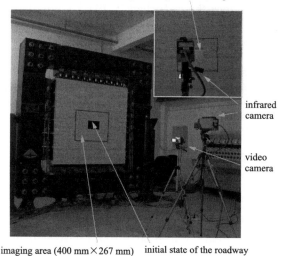

Figure 8.3 Photograph showing the IR imaging site

3. Room temperature influence

Room temperature has a significant influence on the acquired IR image during laboratory test. It was found that if the thermal imaging is conducted in the winter or summer when the indoor and outdoor temperature differences are significant, the raw thermograms are extremely obscure due to the room temperature fluctuation generated from the heat convection or gradient.

In order to control the influence of the room temperature fluctuation on or near exposed surfaces, following measures were taken:

(1) 24 hours earlier prior to the testing, thermal imager as well as all the instruments were brought to the same room with the physical model to balance

the initial temperature difference, so that the detected temperatures are the temperature variation caused by the external loading.

(2) The experiment should be arranged at night in the spring or autumn when there are no strong wind and rain fall. This arrangement is under the consideration that ①there is no need for the use of air conditioner or heater in the room and no strong convection between the indoor and the outdoor; ②at night, the effects of atmospheric and background radiation will be small. This test was conducted at 21:00 PM in September, 2009 at room temperature around 25℃ and the weather was very good.

(3) During the imaging, the unnecessary walking of the laboratory personnel was prohibited to reduce the disturbance. Black-colored curtains were used to cover the background of the detection and windows of the testing room, to prevent the imaging area from influence due to the background radiation.

(4) Image processing techniques were also employed to reduce the room temperature influence. The sequencing filters for noise reduction will be introduced in section 8.4 in the following text.

8.3 Problem statement

Infrared thermography has been applied in the field of engineering rock mechanics for detecting the laboratory specimen sized objects (Luong, 1990, 1995; Wu and Wang, 1998; Wu et al., 2002; Grinzato et al., 2004; Pastor et al., 2008). The smaller the detecting object is, the higher the spatial resolution and contrast. In this case, noise reduction may be the major task for image processing (Gong et al., 2013a).

As stated previously, when detecting the large-scale geological models with the thermal imager working in the passive mode, i. e., without an extra heater for heating the detected objects (He et al., 2009, 2010a, 2010b; He, 2011; Gong et al., 2013), image enhancement should be performed due to the low contrast of the raw thermal sequence.

Reference (He et al., 2009) reported our first experiment on roadway failure in horizontally stratified rock masses. The algorithm (referred to "old algorithm" thereafter) used in the previous study (He et al., 2009) consists of two filters, i. e., *median* filter for removing the pulsation noise and *Canny* filter for sharpening borders and edges of the regions of interest. However, the old algorithm did not work very well in enhancement of the low-contrast thermogram, and the processed images in the former study (He et al., 2009) provide little information about rock behavior.

Major task for the research presented in this chapter is to develop a robust and efficient image enhancement filter, thus develop a new algorithm which incorporates the noise reduction and image enhancement, and at the same time deepen our understanding on the failure mechanisms of the simulated roadway tunnel in the horizontally bedded strata (He et al., 2009, 2010a, 2010b; He, 2011; Gong et al., 2013).

8.4 Image denoising filters

8.4.1 Types of the noise

During the experiment, thermograms are prone to be contaminated by noises from the environmental radiation and electronic components or rotating parts in the testing instruments (Jeng and Woods, 1991; Gonzalez et al., 2005; Tiago et al., 2013). Under the testing condition in the geological model experiments conducted in our institute (He et al., 2009, 2010a, 2010b; He, 2011; Gong et al., 2013), the noises in the raw thermograms include following three types: ① environmental radiation noise; ② impulsive noise (also salt and pepper noise) which is induced by the electronic current in the testing instrument; and ③ additive-periodical noise generated in the rotating parts of the cooling system in the thermal camera (Gong et al., 2013b).

8.4.2 Removing environmental noise

For removing the environmental radiation noise, image subtraction filter could be employed (He et al., 2010a, 2010b; He, 2011). The image subtraction filter is defined by

$$f_k(x,y) = f_k(x,y) - f_0(x,y) \tag{8.1}$$

where, subscript k is the frame index for IR sequence and takes integral values, $k = 0,1,2,\cdots$; $f_0(x,y)$ is image matrix of the first frame in the sequence acquired when the rock was at initial state; x and y are pixel coordinates denoting the pixel number ($x = 1,2,\cdots, M = 160$; $y = 1,2,\cdots, N = 120$; M and N are the maximum coordinate values for x and y respectively), and accordingly, $f_k(x,$

y) is the matrix for the kth frame.

Image matrix $f_0(x,y)$ is in fact the temperature field induced by the environment radiation. Subtraction of the first frame from the following sequence can remove the environmental radiation noises. It is noted that useful information contained in IR images is the incremental temperature field of the deforming rocks, which can also be acquired by the subtraction operation expressed by Eq. (8.1).

8.4.3 Suppression of the impulsive noise

Median filter are proved being efficacious for suppression of the impulsive noise (He et al., 2010a, 2010b; He, 2011; Gong et al., 2013a). The *median* operation is based on ordering the pixels contained in an image area encompassed by the filter, and replacing the central pixel with the ranking result. Given S_{xy} representing the structural element (or mask) centered at the point (x,y) which actually is a matrix of $M \times N$ dimension, $g(x,y)$ representing the filtered image, and $\hat{g}(x,y)$ the noisy image, then the *median* operation is the replacement of the pixel at (x,y) with the median pixel of the neighborhood defined by the mask.

The *median* filter is given by

$$g(x,y) = \underset{(s,t) \in S_{xy}}{\operatorname{median}} \{\hat{g}(s,t)\} \tag{8.2}$$

Details for implementation of the median operation can be found in reference (He et al., 2010a).

8.4.4 Removing the additive-periodical noise

Quality decrease in the captured raw thermograms caused by the additive-periodical noise was found due to the built-in cooling system and infrared sensors (He et al., 2010a, 2010b; He, 2011; Gong et al., 2013a). For removal of the periodical noise, frequency domain filters, such as Butterworth or Gaussian low-pass/highpass filters can be used. In the practical application (Gong et al., 2012; Gong et al., 2013b), Gaussian-high-pass filter (GHPF) was proved being more effective for removing the periodical noise.

A noisy image $g(x,y)$ can be expressed by the model

$$g(x,y) = f(x,y) + \eta(x,y) \tag{8.3}$$

where, $f(x,y)$ is the clean image; $\eta(x,y)$ is the additive-periodical noise. Compute 2-dimensional discrete Fourier transform (DFT) of Eq. (8.3) yields,

$$G(u,v) = F(u,v) + N(u,v) \qquad (8.4)$$

where, u and v are the frequency variables in the horizontal and vertical directions (in unit: Hz).

Let $H(u,v)$ stand for the Fourier transform of the filtering function, then multiply Eq. (8.4) with $H(u,v)$ and assuming that the product of $H(u,v)N(u,v)$ vanishes, then the filtration is implemented by

$$G(u,v) = H(u,v)F(u,v) \qquad (8.5)$$

Perform the inverse Fourier transformation on $G(u,v)$ in Eq. (8.5), i. e., $g(x,y) = \text{DFT}^{-1}[G(u,v)]$, then the cleaned image is recovered from $g(x,y)$. The transfer function $H(u,v)$ for GHPF is defined as (Gonzalez et al., 2005),

$$H(u,v) = 1 - e^{-D^2(u,v)/2D_0} \qquad (8.6)$$

where, D_0 is the cut off frequency of the filter; D is the distance from a point (u,v) to $(M/2, N/2)$ which is the center for an $M \times N$ image, and

$$D(u,v) = [(u-M/2)^2 + (v-N/2)^2]^{1/2} \qquad (8.7)$$

8.5 Morphological enhancement filter

8.5.1 Short review

High or low temperature regions in an IR image are distinguished by borders or edges which are the basis for analysis of the structural changes. First-order derivative or gradient is sensitive to those regions where gray scales change sharply. Image sharpening can be achieved with derivative-based filters such as *Sobel*, *Prewitt*, *Roberts*, and *Canny* filters (Gonzalez et al., 2005).

Image sharpening, however, merely detects the borders or edges and does not alter magnitude of detected pixels. Thus, derivative-based mechanism does not work well with low-contrast images. For example, the IR images in the previous study (He et al., 2009) were enhanced with *Canny* filter and the resultant visual quality is poor. In this section, a morphology-based filter is developed for enhancing low-contrast image.

8.5.2 Fundamentals

The aim for image enhancement is to highlight the image features, i. e., borders, edges and contrast so as to make the image better suited for human visual perception or machine recognition. Mathematical morphology is a widely used methodology for image analysis, smoothing and filtering, segmentation, edge detection, thinning, shape analysis and enhancement.

In addition to the derivative-based filter, some other approaches, such as histogram-based and spatial-based filters, could be used for image enhancement. In the case of low-contrast image with a reduced dynamic scope, however, the morphology-based filters are proved being the most effective for image enhancement (Zeng et al., 2006; Tang et al., 2008; Bai et al., 2011; Gong et al., 2013a).

For the development of the morphological enhancement filter, two basic morphological filters are employed. One is *opening*, $\gamma_B(f)$, where B is the structural element (SE) (also known as template/mask/window) and f the grayscale image. $\gamma_B(f)$ is a non-expanding transformation which filters high-intensity pixels and extracts the minimum in the SE's neighborhood.

Another is *"white" top-hat* (WTH). The WTH transform is made from difference between the initial image and the opened image with respect to a SE B of size λ. The WTH filter is defined as (Soille, 2008).

$$\mathrm{WTH}_B(f) = f - \gamma_B(f) \tag{8.8}$$

According to Eq. (8.8), subtraction of the result of the opening, $\gamma_B(f)$, from an image f will obtain light regions (also foreground) of the image.

8.5.3 Filter development

Major tasks for the filter development include ①extracting image features (foreground or peaks), and ②enlarging the features to certain levels of magnitude. It is seen that the WTH filter extracts image foreground, i.e., all the peaks of the gray scales. If difference between the peaks and valleys is small, the contrast will be low. Hence, a natural idea about enhancement of the low-contrast image could be worked out, i.e., utilizing WTH filter for extracting the peaks in gray scales and then magnifying these peaks.

According to Eq. (8.8), performing WTH transformation on an image, f, yields the results of $f - \gamma_B(f)$ which denotes peaks of the gray scales. For enhancement of the foreground of an image, the peaks in the original image can be magnified by adding the attenuated $[f - \gamma_B(f)]$ by multiplying a factor, $1/\lambda$, i.e., $[f - \gamma_B(f)]/\lambda$, where λ is the size of the SE.

The reason why magnifying the foreground lies in the fact that light regions in an IR image indicate locations where intense compression or friction takes place, providing important information about the progressive development of rock failure. Another consideration is the multi-scale nature of the IR images acquired in rock failure experiments. Rock failure contains multi-scale process including grain-scale deformation, microcrack initiation, propagation and coalescence (Connolly and Copley, 1990). Thus, the size λ of the structuring element should be multi-scale. Accordingly, the added foreground $[f - \gamma_B(f)]/\lambda$ is a function of the size λ, i.e., for small SE, the added foreground has larger magnitude and vice versa.

Following this rule, ideas about enhancement of the low-contrast image based on the principle of magnifying image foreground (MIF) could be formulated as,

$$\text{MIF} = (\text{WTH}_{B_\lambda})/\lambda + f = (f - \gamma_{B_\lambda})/\lambda + f \tag{8.9}$$

where, MIF represents the developed enhancement filter by means of magnifying image foreground; B_λ denotes the multi-scale SE B of size λ; WTH_{B_λ} is the multi-scale white top-hat filter; γ_{B_λ} is the multi-scale opening filter.

The kernel for implementation of the *MIF* filter is the multi-scale operations by using the filter γ_{B_λ} which is defined by,

$$\gamma_{B_\lambda} = \sum_{\lambda=1}^{n} \gamma_{B_\lambda} \tag{8.10}$$

where, λ is the widths for a square-formed SE or radius for a disc-formed SE, and λ takes discrete value and increases by odd numbers relation $\lambda = 1, 3, \cdots; n = 2k+1$, $k = 1, 2, 3, \cdots$; k is the scaling factor. It is seen from Eq. (8.9) that the filter MIF magnifies the foreground $(f - \gamma_{B_\lambda})$ by $1/\lambda$ and keeps the background unchanged.

The background of infrared images obtained in rock mechanics experiment denotes the temperature reduction effects due to relaxation, tensile fracture, etc. (Gong et al., 2013b). The use of $1/\lambda$ as attenuation factor in Eq. (8.9) can ensure that the gray scales for small-scale objects (corresponding small-scale ruptures) are magnified to a relatively higher levels, and the gray scales for large-scale objects (corresponding to macroscopic fractures) are magnified with moderate levels.

8.5.4 Multi-scale SE

1. Multi-scale approach

A structural element is an elementary tool of a given simple geometry used to transform the image (Soille, 2008). It is a part of the picture with one central point (i. e., template origin) distinguished. When the template origin is placed on a pixel in an image under study, SE determines a neighborhood area. When the SE is moved over the whole image, for each point, a coincidence analysis of the points of the picture and the points of the SE is carried out.

For all points of the picture, the compatibility of the pixel configuration in the closest neighborhood of the tested point within the SE is checked. In the case of conformity, the earlier defined operation is performed on the tested point (Chermant and Coster, 1994). Considering the fact that most of the detecting objects are rectangular-shaped in our test (He et al., 2009), a square-formed SE was employed.

2. Multi-scale template

For design of B_λ, estimation of the spatial resolution that detects the smallest fracture is required. In the test (He et al., 2009), the imaging area is 400 mm×367 mm and the IR images were stored by 120×160 pixels. Accordingly spatial resolution for the IR image is 2.8 mm/pixel which evaluated as the square root of the imaging area divided by the total image pixels. Namely, one pixel occupies a square area with 2.8 mm side length.

Because thickness of the elementary slab used for construction of the model rock strata is 10 mm (see section 8.2), a slab-scale fracture could be detected by using one-pixel sized structural element. At the same time, the deforming object in the imaging area is the roadway section with a face area of 200 mm (height) × 250 mm (width), the corresponding size for SE, λ, should not be less than 90 pixels.

3. Scaling factor

Values for λ used in this research are ranged from one pixel to 95 pixels, equivalent to a series of square-formed imaging areas with side lengths from 2.8-266 mm. Accordingly, the side lengths for the square-formed SEs are $\lambda = 1, 3,$

..., $n = 2k + 1$; $\lambda_{max} = 95$ pixels ($k = 1, 2, 3, \cdots, 47$).

The SE's size λ was determined under the following two considerations: ①the side length for SE should be sufficiently small capable of detecting the small scale cracks; thus the minimum size of SE is one pixel; ②the side length for SE should cover the roadway section, and $\lambda = 95$ pixels meets this requirement; ③the selection of $\lambda = 95$ is to ensure that the progressively converging roadway section under stressing could be detected.

8.6 Image processing

8.6.1 Algorithm and image analysis rules

For processing the IR images in the previous study (He et al., 2009), following algorithm are proposed and listed in the processing sequence:

(1) Subtraction of the first frame from the follow-up IR images for eliminating the background radiation noises;

(2) Performing the *median* filter for suppressing the impulsive noise;

(3) Performing the Gaussian-high-pass filter, i.e., GHPF, for removing the additive-periodical noise;

(4) Finally, performing the newly developed morphological enhancement filter, MIF, for enhancement of the denoised IR images.

The above image processing operations were realized in the MATLAB 8.0 platform based on the Image Processing Toolbox (IPT) functions in the Matlab macro code.

Infrared image represents the precursory information about rock fracture by pseudo-colors, borders and edges and light or dark regions which are the different temperature distribution zones.

For facilitating comprehension of the IR images, following well-established rules can be used (Gong et al., 2013a):

(1) hot or positive colors stand for high-level temperatures and cool or negative colors for low-level temperatures;

(2) high temperature indicates high stress level due to friction, shear or stress concentration, and low temperature indicates low stress level due to tensile cracking, stress release or unloading;

(3) temperature distribution scale in the light or dark regions may indicate the scale of the rock damage and localized temperature distribution represents

plastic deformation or permanent damage;

(4) the borders and edges that separate hot- and cool- colored zones illustrate the modes of rock behavior.

These rules are used in comparison of the visual quality between the IR sequence processed using the old algorithm in the previous work (He et al., 2009) and the same sequence processed using the new algorithm developed in this research.

8.6.2 Assessment of imaging processing effect

According to the stress path in Figure 8.2, some typical frames in the acquired infrared sequence are selected for manifesting the rock responses at different stress states including the hydrostatic loading phase marked by A1-A5 and non-hydrostatic loading phase marked by B1-B6, in accordance with the IR sequence in the previous study (He et al., 2009).

Comparison of the image processing effect for frame 010 at stress state A1 ($\sigma_1 = \sigma_2 = 0.8$ MPa) is shown in Figure 8.4. Figure 8.4a is the raw IR image of frame 010 acquired in the previous study (He et al., 2009). Figure 8.4b is the frame 010 processed by using the old algorithm in He et al. (2009). Figure 8.4c is the same frame in He et al. (2009) processed by the new algorithm. It is seen that the raw frame 010 is noisy and dim in contrast.

Only a rectangular shaped tunnel section could be observed. In the frame 010 processed by the old algorithm, only two thin belts of high temperature, representing the stress concentration, on the two side walls of the tunnel section could be observed. It was known that the stress redistribution will take place at the excavation boundary and vanish progressively into the far-field region (Brady and Brown, 2005). Therefore, the old image in He et al. (2009) does not provide correct information on rock mass response.

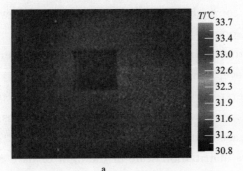

a

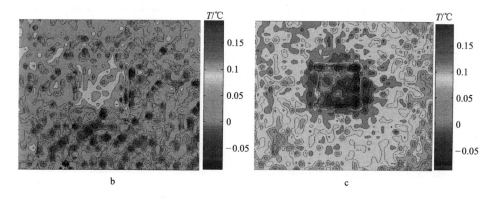

Figure 8.4 Comparison of the image processing effect

a. frame 010 acquired at stress state A1 ($\sigma_1 = \sigma_2 = 0.8$ MPa); b. frame 010 processed by using the old algorithm in the previous study (He et al., 2009); c. frame 010 processed by the new algorithm proposed in this paper

In the frame 010 (Figure 8.4c) processed by the new algorithm, high temperature distributes in the near-face region roughly symmetric to the center of the roadway section, corresponding to the low-level hydrostatic stress state A1. At the tunnel boundaries including the roof, floor and two side walls, high temperature indicates the stress concentration, and the temperature tends to decrease to the same level as that in the far-field regions.

The new infrared image 010 represents the stress redistribution of the simulated tunnel boundaries in conformity with the well established principles as stated in reference (Brady and Brown, 2005). Comparison to the old image of the same frame in He et al. (2009) demonstrates a marked improvement using the new algorithm and image enhancement filter, MIF. It is worth mentioning that compared with the morphological enhancement filter, κ_n, developed in reference (Gong et al., 2013a), the new filter, MIF, has a simple structure and higher efficiency, as well as less disturbance to the image background. Within the tunnel section, temperature distribution is the same as the room temperature, lower than that in the stressed rock mass. It is seen that the tunnel section is not clear in the old image (Figure 8.4b) and very clear in the new image (Figure 8.4c).

8.6.3 Rock response at hydrostatic stress state

Figure 8.5 to Figure 8.7 show infrared images and photographs corresponding to the hydrostatic stress states from A1 to A5. The images in Figure 8.5 are processed with the old algorithm proposed in the previous work (He et al., 2009) (referred to as "old IR images").

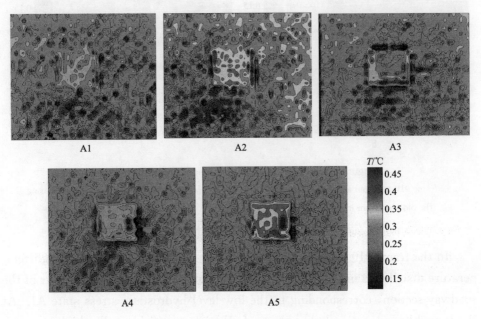

Figure 8.5 Old IR images in the previous study (He et al., 2009); stress states
A1 (0.8, 0.8 MPa), A2 (0.8, 0.8 MPa), A3 (0.8, 0.8 MPa), A4 (1.4, 1.4 MPa), and
A5 (1.4, 1.4 MPa)

Infrared images in Figure 8.6 are processed with the new algorithm developed in this paper (referred to as "new IR images"), and images in Figure 8.7 are video screenshots captured simultaneously with infrared imaging (referred to as "photographs"). Note that the black rectangle in the photographs is actually a 400 mm×300 mm red-colored frame of plastic tape indicating the infrared imaging zone. The applied stresses (σ_1, σ_2) for each of the stress states are also given in the captions in Figure 8.5 to Figure 8.7.

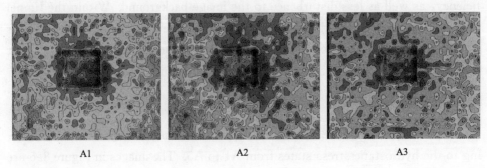

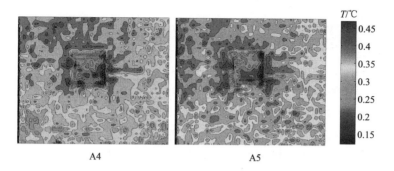

Figure 8.6 New IR images; stress states
A1 (0.8, 0.8 MPa), A2 (0.8, 0.8 MPa), A3 (0.8, 0.8 MPa), A4 (1.4, 1.4 MPa), and A5 (1.4, 1.4 MPa)

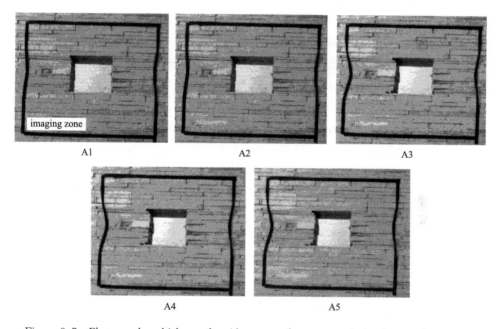

Figure 8.7 Photographs which are the video screenshots captured simultaneously with the infrared imaging; the black rectangle indicates the infrared imaging zone; stress states A1 (0.8, 0.8 MPa), A2 (0.8, 0.8 MPa), A3 (0.8, 0.8 MPa), A4 (1.4, 1.4 MPa), and A5 (1.4, 1.4 MPa)

By comparison with new IR images (Figure 8.6), it is seen that major deficiencies for the old IR images (Figure 8.5) include:

(1) information on the stress redistribution is less;

(2) stress concentration occurs only at the excavation boundary, which is

not the truth;

(3) boundary of the roadway tunnel is not shown clearly;

(4) an inversion of warm colors to the interior of the opening, suggesting either poor processing or influence of room temperature fluctuations in the opening.

On the contrary, new IR images in Figure 8.6 provide different modes of the temperature distribution representing rock responses. New IR images will be analyzed and compared to the photographs in the following text.

As analyzed previously, new infrared image A1, under the stress state (0.8, 0.8 MPa), shows that stress redistribution is roughly isotropic; deformation of the tunnel section is small and roughly symmetric; hot colors on the two side walls indicates stress concentration caused by compression and static friction on the bedding planes; hot colors on the roof represents the temperature rise as a result of the overburden-induced compression, and hot colors on the floor denotes the temperature rise due to squeezing caused by the inward displacement of the two side walls. Rock behaviors illustrated in the new IR image could not be observed in the photograph A1.

It is seen in the new infrared image A2, under (0.8, 0.8 MPa), that: ① the overlying strata and two side walls were stressed heavily, represented by darker hot colors with larger extent; ② the two side walls were displaced further toward the opening. These phenomena could also be seen in the photograph A2.

Prominent feature seen in the new IR image A3, under (0.8, 0.8 MPa), is the emergence of the stratification-paralleled high temperature belts adjacent to the right-side wall representing strong static friction on the bedding planes. In photograph A3, further displacement of the two side walls toward the opening were observed. However, the static friction between the rock layers could not be seen clearly in the photograph.

New IR image A4 under, (1.4, 1.4 MPa), illustrates that the roadway section converges further as a result of displacement of the two side walls. Floor heave was seen clearer in the IR image and evidenced by the photographs A4.

Dramatic feature for the new IR image A5, under (1.4, 1.4 MPa), is that the stratification-paralleled temperature belts adjacent to the right-side wall became distinct, i. e., below the hot-colored belt, there is a cool-colored belt, indicating the transition from the static friction to dynamic friction. Phenomena of the friction between rock layers can not be observed clearly in the photograph A5.

Analyses and comparison of the image sets in Figure 8.5 to Figure 8.7 demonstrate the fact that when roadway deformation is small under the hydrostatic loading at low stress level, new IR images provide different modes of the precursory information represented by temperature distributions, i. e. ,

(1) scattering distribution indicates the elastic response;

(2) localized distribution indicates rock friction, stress concentration or plastic damage;

(3) the shape, scale, and temperature level of the localized temperature zones determine the rock mass failure mechanism;

(4) under hydrostatic loading, temperature distributions were not symmetric and but anisotropic with increasing heterogeneity.

8.6.4 Rock response at unbalanced stress state

Figure 8.8 to Figure 8.10 show IR images and photographs corresponding to the unbalanced stress states from B2 to B6. The old IR images in the previous study (He et al., 2009) are shown in Figure 8.8, new IR images in Figure 8.9, and photographs in Figure 8.10. The related stress states (σ_1, σ_2) are also given in the captions for Figure 8.8 to Figure 8.10.

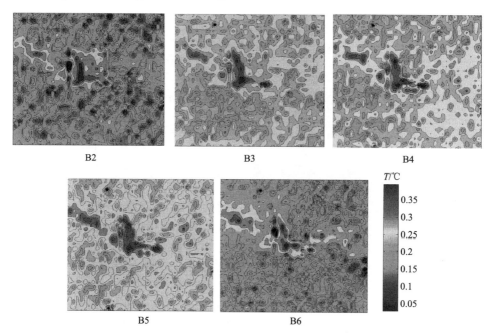

Figure 8.8 Old IR images in He et al. (2009); stress states
B2 (2.5, 1.4 MPa), B3 (3, 1.4 MPa), B4 (4, 1.4 MPa), B5 (5, 1.4 MPa), and B6 (6, 2 MPa)

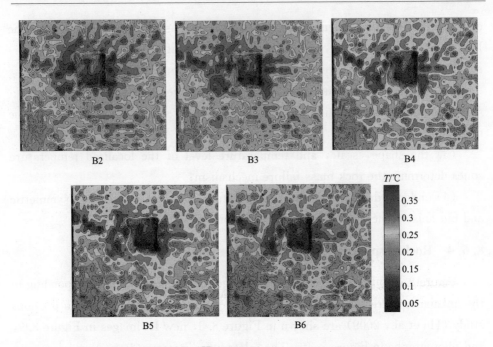

Figure 8.9 New IR images; stress states; stress states
B2 (2.5, 1.4 MPa), B3 (3, 1.4 MPa), B4 (4, 1.4 MPa), B5 (5, 1.4 MPa), and B6 (6, 2 MPa)

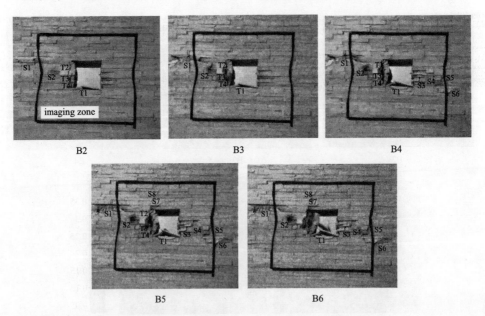

Figure 8.10 Photographs which are the video screenshots captured simultaneously with the infrared imaging; the black rectangle indicates the infrared imaging zone; stress states; stress states
B2(2.5, 1.4 MPa), B3(3, 1.4 MPa), B4(4, 1.4 MPa), B5(5, 1.4 MPa), and B6(6, 2 MPa)

Deficiencies for the old images lie in the insufficient information about fracture development. Primary conclusion drawn from analysis of the old images is the oblique macroscopic fracture on the left side of the imaging plane and details are lacking. It is seen from Figure 8.9 that the new IR images provide different modes of the precursory information about rock response. Analysis of the new IR images and comparison to the photographs are given in the following text.

It is seen from the new IR image B2 under (2.5, 1.4 MPa) that:

(1) hot-colored belts parallel to the bedding planes represents the intense rock friction adjacent to the right-side wall;

(2) temperature localization on the left-side wall and upper-left corner signals the imminent fracture event;

(3) several hot-colored zones on the left half plane along an oblique line indicates fracture initiation;

(4) floor heave and marked convergence of the roadway section could be observed.

It is seen from the photograph B2 that:

(1) rock friction on the right-half plane are evidenced very well;

(2) an opened vertical fracture was seen clearly adjacent to the left-side wall (could not be seen in the IR image);

(3) damage on the upper-left corner could hardly be seen (but seen clearly in the IR image);

(4) fracture initiation along the oblique line on the left-half plane could be seen but not so clear compared with the IR image.

By comparison of the IR image with photograph at the stress state B2, it can be understood that:

(1) IR image is capable of providing precursory warnings on rock failure when the fracture development involves shear, friction and compression, having strong thermal-mechanical coupling effect;

(2) IR image is incapable of detecting geometrical details of the opened fracture having a weak thermal-mechanical coupling effect.

It is seen from the new IR image B3 under (3, 1.4 MPa), one can understand that:

(1) high temperature zones on the left-half plane coalesced into an oblique belt across the bedding planes denoting a shear fracture at macroscopic scale;

(2) the connected hot- and cool-colored belts on the right-half plane indicates

the frictional sliding on the bedding planes;

(3) significant displacement of the right-side wall was seen.

In comparison to photograph B3, it is observed that:

(1) the oblique shear fracture on the left-half plane was seen clearly;

(2) the frictional sliding was evidenced;

(3) a slab-like rock block separated by the opened vertical fracture adjacent to the right-side wall was seen clearly; however, the IR image could not detect the geometrical structure of the opened fracture as a result of the weak thermal-mechanical coupling;

(4) for the same reason, the marked floor heave was not seen clearly in the IR image.

Major features seen in the IR image and photograph B4 under (4, 1.4 MPa) include:

(1) further displacement of the separated block adjacent to the left-side wall was seen clearly in the photograph, which could also be seen in the IR image, but not so clear;

(2) floor heave due to uplift of the cracked rock layer was not shown in the IR image;

(3) the friction-induced fracture on the right-half plane was seen clearly both in the IR image and photograph.

A marked event seen in the new IR image B5 under (5, 1.4 MPa) is that a coalesced vertical fracture on the upper-left corner (also shown in the photograph). The event was predicted by localization of the high temperature on the upper-left corner in the IR images B2 to B5. It is noted that the temperature distribution under unbalanced loading showed stronger anisotropy and heterogeneity than that under hydrostatic loading (see Figure 8.8 to Figure 8.10).

In the new IR image B6 under (6, 2 MPa): ①significant convergence of the tunnel section due to displacement of the left-side wall was seen clearly, correspondingly, partial failure of the rock block due to compression of the bending roof was shown in the photograph; compressive deformation generates heat which can be detected by the thermography; this is why the displaced rock block was represented in the IR image; ② the floor heave due to squeezing by the inward displacement of the two side walls can also be detected for the same reason; as a result, the floor heave was seen both in the IR image and photograph.

From the analyses and comparison, capabilities of the new IR images for representing rock mass response could be summarized as:

(1) Initiation, propagation and coalescence of the compressive, frictional and shear fractures can generate a lot of heat due to the intense thermal-mechanical coupling. As a result, the IR image can provide forewarnings on rock response by the temperature distributions;

(2) In comparison to the photographs, fracture events predicted by the IR image are well evidenced by the video photographs, i. e., the forewarning messages represented by the temperature distributions in the new IR images are reliable and accurate;

(3) Progressive development of the opened fractures produces less heat due to weak thermal-mechanical coupling effect. In this case, the IR image can hardly detect geometrical details of the opened fractures and cracks;

(4) Compared to the old algorithm in the previous study (He et al., 2009), the morphological filter developed in this study magnifies the foreground with a scaling factor inversely proportional to the gray levels and without disturbance to the background, so that make the small changes in the temperature field understandable while retains the boundary of the tunnel section.

New IR images outperform the old ones in He et al. (2009) in the following aspects:

(1) Representing clearly the stress redistribution and concentration for the simulated roadway tunnel under the hydrostatic loading when the stress level is low;

(2) Providing accurate forewarning messages for the critical static/dynamic frictions and the resultant frictional sliding induced failure;

(3) Providing different modes of the precursory information about the initiation, propagation and coalescence of the cracks and eventual failure;

(4) Representing clearly the structural changes of the tunnel section including the inward displacement of the two side walls, roof bending and subsidence, as well as the floor heave with image features.

8.7 Characterization of new IR images

8.7.1 Mission and rule

At unbalanced stress states, the roadway will be prone to failure. Therefore, more new IR images corresponding to the stress states B1-B6 will be analyzed in this section, for exploring the non-linear rock responses. Convention for the image label is, for example, B1-1 standing for a frame acquired after image B1, B1-2 for a frame acquired after B1-1, at the same stress state as B1, and so on.

Figure 8.11 and Figure 8.12 show the IR sequence and video photographs under the stress state B1, (1.8, 1.4 MPa) with lateral pressure coefficient λ equal to 0.78. As seen in image B1, ①roof and two-side walls were stressed heavily; ②stratification-paralleled hot- and cool-colored temperature belts on the left-side plane represent the static and dynamic frictions respectively; ③crack initiation is indicated by several isolated small-scale hot-colored regions lined obliquely on the left-side plane.

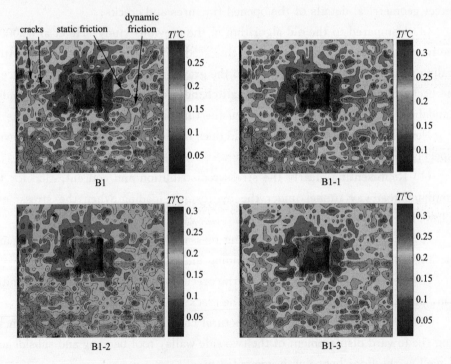

Figure 8.11 IR images at the stress state B1, $\sigma_1 = 1.8$ MPa, $\sigma_2 = 1.4$ MPa and $\lambda = 0.78$; the black rectangle indicates the infrared imaging zone; the convention for the image label is: B1-1 standing for a frame acquired after image B1, B1-2 for a frame acquired after B1-1, and B1-3 standing for a frame acquired after B1-2, at the same stress state as B1

Chapter 8 Overloaded tunnel in horizontal strata · 255 ·

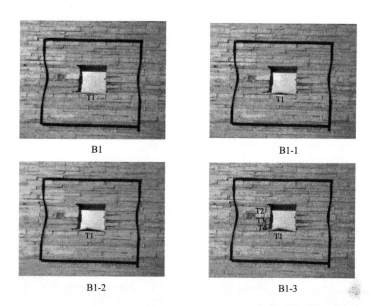

Figure 8.12 Photographs at the stress state B1, $\sigma_1 = 1.8$ MPa, $\sigma_2 = 1.4$ MPa and $\lambda = 0.78$; the black rectangle indicates the infrared imaging zone; the convention for the image label is: B1-1 standing for a frame acquired after image B1, B1-2 for a frame acquired after B1-1, and B1-3 standing for a frame acquired after B1-2, at the same stress state as B1

8.7.2 Loading case B1

Images B1-1, B1-2 and B1-3 illustrate evolution of these fractures. By comparison to the photographs, it is seen that:

(1) IR images well represent the failure mechanism by temperature distribution patterns such as the static and dynamic frictions, stress concentration, and crack initiation; while these precursory signs could hardly be observed in the photographs when the deformation was small;

(2) The floor heave and displacement of the two-side walls were observed both in the IR sequence and photographs;

(3) Photograph B1-3 shows that a slab-shaped rock block was separated by an opened vertical fracture on the left-side wall; the IR sequence can not detect the details of the opened fracture but provide precursory information on the fracturing event.

8.7.3 Loading case B2

Figure 8.13 and Figure 8.14 show the IR sequence and video photographs at stress state B2, (2.5, 1.4 MPa) with $\lambda = 0.56$. As analyzed in section 8.6.4, the prominent features for the new IR image B2 include:

(1) crack propagation along the oblique line on the left-half plane represented by growth of the high temperature regions;

(2) the static/dynamic frictions represented by the high/low temperature belts adjacent to the right-side wall. Images B2-1, B2-2 and B2-3 illustrate coalescence of the cracks which finally merged into an oblique shear fracture on the left-side plane.

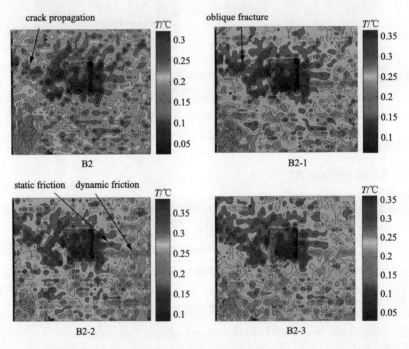

Figure 8.13 IR images at the stress state B2, $\sigma_1 = 2.5$ MPa, $\sigma_2 = 1.4$ MPa and $\lambda = 0.56$; the black rectangle indicates the infrared imaging zone; B2-1, B2-2 and B2-3 denote the successively acquired frames at the same stress state as B2

From the comparison with the photographs, it is seen that:

(1) Precursory information in the IR sequence about coalescence of the oblique shear fracture is accurate, which was observed clearly in the photographs in the later phase;

Chapter 8 Overloaded tunnel in horizontal strata · 257 ·

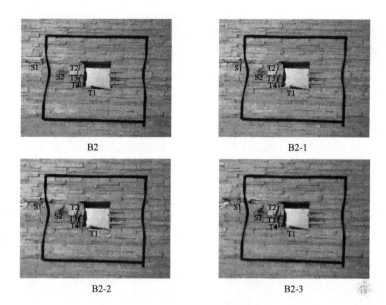

Figure 8.14 Photographs at the stress state B2, $\sigma_1 = 2.5$ MPa, $\sigma_2 = 1.4$ MPa and $\lambda = 0.56$; the black rectangle indicates the infrared imaging zone; B2-1, B2-2 and B2-3 denote the successively acquired frames at the same stress state as B2

(2) Frictional sliding failure mechanism on the right-side plane represented by the IR sequence can also be observed clearly in the photographs in the later phase;

(3) Floor heave due to uplift of the cracked rock layer could be observed both in the IR images and photographs; details of the already-cracked rock layers can only be seen in the photographs;

(4) Structure of the rock block separated by the vertical fracture on the left-side wall in the photograph can not be represented as the opened crack produces less heat.

8.7.4 Loading case B3

Figure 8.15 and Figure 8.16 show the IR sequence and video photographs at stress state B3, (3.0, 1.4 MPa) with $\lambda = 0.47$. Analysis and comparisons showed that:

(1) The oblique shear fracture across the bedding planes on the left-half plane was observed clearly in the IR images and photographs;

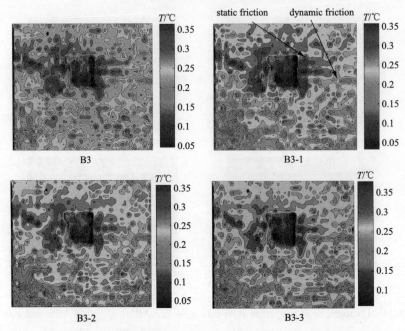

Figure 8.15 IR images at the stress state B3, $\sigma_1 = 3.0$ MPa, $\sigma_2 = 1.4$ MPa and $\lambda = 0.47$; the black rectangle indicates the infrared imaging zone; B3-1, B3-2 and B3-3 denote the successively acquired frames at the same stress state as B3

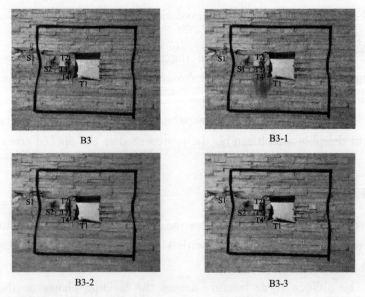

Figure 8.16 Photographs at the stress state B3, $\sigma_1 = 3.0$ MPa, $\sigma_2 = 1.4$ MPa and $\lambda = 0.47$; the black rectangle indicates the infrared imaging zone; B3-1, B3-2 and B3-3 denote the successively acquired frames at the same stress state as B3

(2) Marked frictional sliding between rock layers adjacent to the right-side wall was seen clearly in the IR images and photographs;

(3) Progressive convergence of the roadway section due to the inward displacement of the two side walls, roof bending and floor heave were observed both in the IR images and photographs.

8.7.5 Loading case B4

Figure 8.17 and Figure 8.18 show the IR sequence and video photographs at stress state B4, (4.0, 1.4 MPa) with $\lambda = 0.35$. It is seen that:

(1) High temperature localization on the upper left corner signals a forewarning on the crack propagation as seen in the image B4; this forewarning could hardly be observed in the photograph B4;

(2) Bending of the roof and further propagation of the cracks on the upper left corner were well represented by the temperature localization in IR image B4-1;

(3) In the photograph B4-2, the cracks on the upper left corner coalesced

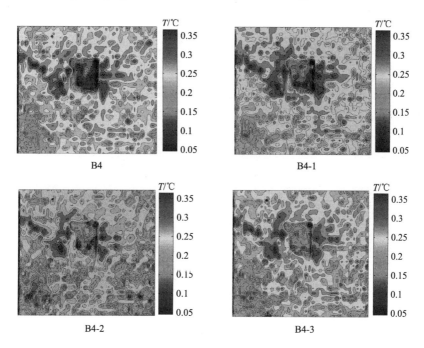

Figure 8.17 IR images at the stress state B4, $\sigma_1 = 4.0$ MPa, $\sigma_2 = 1.4$ MPa and $\lambda = 0.35$; the black rectangle indicates the infrared imaging zone; B4-1, B4-2 and B4-3 denote the successively acquired frames at the same stress state as B4

into a shear fracture almost perpendicular to the bedding planes (vertical); marked roof subsidence, due to cutting off the immediate roof by the vertical fracture, was seen clearly; these events were predicted and represented by the IR sequence B4, B4-1 and B4-3;

(4) In the photograph B4-3, further development of the vertical fracture on the upper-left corner was seen, and hence the increased subsidence of the immediate roof; this vertical fracture belongs to type II, thus represented by the temperature localization in the IR image B4-3;

(5) Block displacement toward the opening was represented by the IR sequence B4; the subsidence of the immediate roof will exert an axial load on the block and the resultant buckling deformation will produce heat; as a result, the convergence of the section caused by the block buckling was detected by the IR images;

(6) Floor heave as a result of the uplift of the cracked rock layers was represented by the IR sequence B4; this uplift is due to the squeezing of the rock layers by the inward displacement of the two side walls.

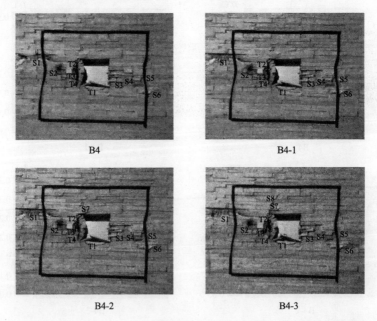

Figure 8.18 Photographs at the stress state B4, $\sigma_1 = 4.0$ MPa, $\sigma_2 = 1.4$ MPa and $\lambda = 0.35$; the black rectangle indicates the infrared imaging zone; B4-1, B4-2 and B4-3 denote the successively acquired frames at the same stress state as B4

8.7.6 Loading case B5

Figure 8.19 and Figure 8.20 shows the IR sequence and video photographs at stress state B5, (5.0, 1.4 MPa) with $\lambda = 0.28$ (the smallest lateral coefficient). From the analyses of the IR sequence B5 and comparison to the photographs, it is seen that:

(1) Rock failure controlled by the oblique fracture on the left-half plane was represented by the high temperature localization in the IR sequence and evidenced by the photographs;

(2) The stratification-paralleled slippage indicated by the horizontal temperature belts in the IR sequence on the right-half plane was also observed in the photographs;

(3) Progressive development of the vertical shear fracture on the upper-left corner and the related roof subsidence were predicted by the IR sequence B3 and B4 and B5, and seen clearly in the photographs;

(4) Floor heave was seen both in the IR sequence and photographs.

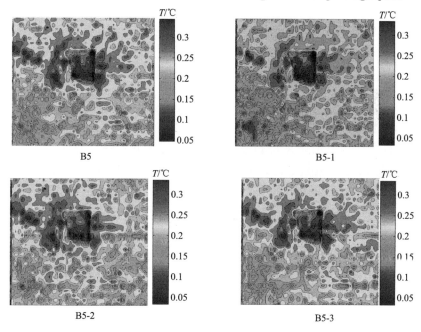

Figure 8.19 IR images at the stress state B5, $\sigma_1 = 5.0$ MPa, $\sigma_2 = 1.4$ MPa and $\lambda = 0.28$; the black rectangle indicates the infrared imaging zone; B5-1, B5-2 and B5-3 denote the successively acquired frames at the same stress state as B5

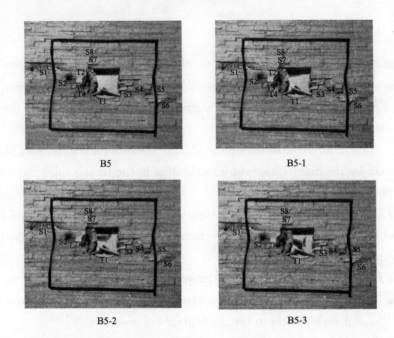

Figure 8.20 Photographs at the stress state B5, $\sigma_1 = 5.0$ MPa, $\sigma_2 = 1.4$ MPa and $\lambda = 0.28$; the black rectangle indicates the infrared imaging zone; B5-1, B5-2 and B5-3 denote the successively acquired frames at the same stress state as B5

8.7.7 Loading case B6

Figure 8.21 and Figure 8.22 show the IR images and video photographs at stress state B6, (6, 2 MPa) with $\lambda = 0.33$ (the highest vertical load). From the analyses (also see section 8.6.4) and comparisons, it is seen that:

(1) Significant subsidence of the roof, partial failure of the block, floor heave and frictional sliding of rock layers on the right-side wall were seen in detail in the photograph B6, also represented by the temperature distribution in the IR image B6;

(2) Collapse of the rock block on the left-side wall was seen in the photograph B6-1; the significant temperature reduction as a result of the relaxation due to the collapse was well represented by the IR image B6-1;

(3) Distortion of the tunnel section due to the complete failure of the roadway was represented by the IR sequence B6-2 and B6-3 and seen clearly in the photographs B6-2 and B6-3; note that the floor defined by the hot-colored temperature belt in the IR images B6-2 and B6-3 does not represents the fractured

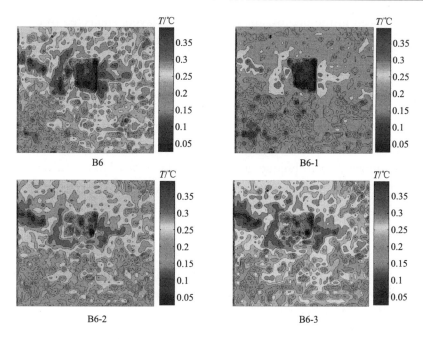

Figure 8.21　IR images at the stress state B6, $\sigma_1 = 6$ MPa, $\sigma_2 = 2$ MPa and $\lambda = 0.33$; the black rectangle indicates the infrared imaging zone; B6-1, B6-2 and B6-3 denote the successively acquired frames at the same stress state as B6

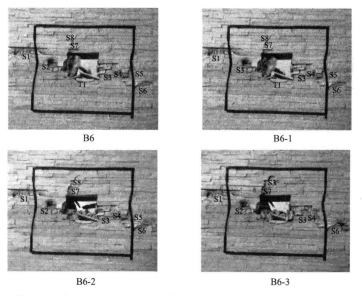

Figure 8.22　Photographs at the stress state B6, $\sigma_1 = 6$ MPa, $\sigma_2 = 2$ MPa and $\lambda = 0.33$; the black rectangle indicates the infrared imaging zone; B6-1, B6-2 and B6-3 denote the successively acquired frames at the same stress state as B6

rock layers seen in the photographs but the rock layers subjected to compressive load exerted by convergence of the two side walls.

8.7.8 Discussion

The presented research is an extension of previous work reported in reference (He et al., 2009). Comparison to previous processing of the same images in earlier work (He et al., 2009) illustrates the significant improvement using the new image processing technique developed in this study. Major advances of this research in comparison to previously published studies (He et al., 2009; Gong et al., 2013a and b) on image processing and analysis of the rock mass response involve the following aspects:

(1) New IR images provide different modes of precursory information on the stress redistribution including crack initiation, propagation and coalescence, static/dynamic friction on the bedding planes and shear or frictional fracture development in different locations of the surrounding rock mass; whereas, old IR images in He et al. (2009) provides less information on the stress redistribution both in modes and locations;

(2) The morphological enhancement filter, MIF, developed in this study, magnifies the foreground of the image with a scaling factor inversely proportional to grey scales of the foreground without any disturbance to the background. The morphological filter proposed in the study (Gong et al., 2013a) magnifies the foreground and alters the background at the same time. Image processing and analysis in this study showed that:

(1) The new technique is robust and efficacious to reduce noises, eliminate the room temperature fluctuations and enhance the image features, and thus make small changes in the temperature field understandable;

(2) Cross-sectional geometry of the simulated tunnel is the image background having the same temperature distribution as the room temperature; as compared to the old image in He et al. (2009), new image represents the tunnel section clearly in the deformation process.

(3) Analyses of the typical new IR images acquired in loading of the roadway tunnel and comparison to the corresponding photographs demonstrate the fact that the infrared forewarning messages represented by the temperature distribution patterns are reliable and accurate.

(4) Different modes of the precursory signs in the new IR images, depicted

by the image features, e. g. the scattering distribution, localized distribution, shapes, levels, borders and edges of the high and low temperature zones, determine the rock failure mechanisms.

Infrared thermography detects phenomena with thermal-mechanical coupling such as friction, shear or compression involved in the deformed rock masses (Luong, 1990; Pastor et al., 2008). However, the thermal camera usually has much less pixels than the visible light cameras. Thus IR image could hardly represent geometrical details of the crack development. At the same time, quantitative analysis of the stress/strain from temperature field has encountered great challenges in de-coupling the energy or heat consumed in different processes when the rock is loaded beyond its elastic limit (Luong, 1990; Gong et al., 2013a). Therefore, typical measurement methods such as deformation monitoring, acoustic emission, etc. are indispensable in the rock mechanics experiment, and the thermography could offer a remote, accurate and fast tool providing a primary or complementary basis for rock mass response analysis.

The thermography TVS-8100 MKⅡ used in our tests (He et al., 2009, 2010a, 2010b; He, 2011; Gong et al., 2013a) is designed specially for the indoor test and not suited for working under strong-convection, vibration, humidity and foggy environment. Up to now, numerous models of the hand-held infrared camera were developed specially for the in-situ tests (Spampinato et al., 2011). The above discussion about superiority of the thermal imaging is on the basis of laboratory condition where room temperature and testing environment could be controlled.

Potential limitations of the thermal imaging in real-life applications, e. g. in engineering rock projects, will be the testing environment. The thermal imager detects temperature variations on the surface in view. As discussed in section 8. 2. 3, environmental radiation has a significant influence on the thermal profile. In the field cases where, though the surface of the rock mass are visible, ventilation and heat from the machinery is overwhelming, real rock response could not be represented by the acquired thermal sequence.

Even if no ventilation and heat from the machinery, the quality for the IR images acquired from the in-situ detection may be much lower than that in the laboratory test because: ① the hand-held infrared camera usually has lower temperature accuracy than that for the indoor thermal imagers, and ② larger imaging zone will reduce the spatial resolution of the image. Therefore, in-situ applications in the practical engineering mechanics projects, thermal imaging could only be used as rough estimation of temperature changes related to the potential danger areas.

It is worth mentioning that, in recent decades, the technology for spaceborne and satellite based infrared cameras has advanced significantly, offering remote sensing applications such as the volcano observations (Spampinato et al., 2011). Theory and techniques for analysis of these air-based remote sensing may be very different from the thermal imaging involved in this document. Thus the related discussions are beyond the scope of the author's knowledge.

8.7.9 Summary

An image enhancement filter, MIF, was developed based on white top-hat transformation and principle of magnifying image foreground. A squared-formed structuring element was used with a multi-scale side length λ having values ranging from the smallest detectable crack to the largest object. Gray-scale peaks, i.e., the foreground of the infrared image are multiplied by a factor, $1/\lambda$, for magnifying small object to a larger magnitude while the larger one to a moderate level.

In order to process the noisy and low-contrast thermograms, new image processing algorithm was proposed which consist of the following operations: *image subtraction* for removal of the environmental radiation noise, *median* filter for reduction of the pulsation noise, Gaussian-high-pass filter, GHPF, for removing the additive-periodical noise, and multi-scale morphological enhancement filter, MIF, for enhancement the low-contrast infrared image.

Rock mass response analysis using the new infrared images was performed. Comparison to the previous processing of the same images in earlier work (He et al., 2009) and to the photographs illustrates the significant improvement using sequence of the noise-reduction filters and image enhancement filter. The forewarnings on rock failure represented by the new image is accurate, and deeper understanding of the failure mechanisms on the simulated tunnel in horizontally bedded rock masses was achieved.

References

Bai X Z, Zhou F G, Xue B D. 2011. Infrared image enhancement through contrast enhancement by using multiscale new top-hat transform. Infrared Phys. Technol., 54(2):61-69.

Brady B H G, Brown E T. 2005. Rock Mechanics for Underground Mining. New York: Kluwer Academic Publishers.

Chermant J L, Coster M. 1994. Role of mathematical morphology in filtering, segmentation and analysis. Acta Stereol, 13:125-36.

Connolly M, Copley D. 1990. Thermographic inspection of composite material. Materials Evaluation, 48(12):1461-1463.

Eberhardt E, Stead D, Stimpson B, et al. 1997. Changes in acoustic event properties with progressive fracture damage. International Journal of Rock Mechanics and Sciences & Geomechanics Abstracts, 34(3-4):633.

Eberhardt E, Stead D, Stimpson B, et al. 1998. Identifying crack initiation and propagation thresholds in brittle rock. Canadian Geotechnical J., 35(2):222-233.

Fortsakis P, Nikas K, Marinos V, et al. 2012. Anisotropic behavior of stratified rock masses in tunneling. Eng. Geo., 141-142:74-83.

Gong W L, Gong Y X, Long A F. 2013a. Multi-filter analysis of infrared images from the excavation experiment in horizontally stratified rock. Infrared Phys. Technol., 56:57-68.

Gong W L, Wang J, Gong Y X, et al. 2013b. Thermography analysis of a roadway excavation experiment in 60° inclined stratified rocks. International Journal of Rock Mechanics and Sciences, 60:134-147.

Gong Y X, Long A F, Gong W L, et al. 2012. Infrared thermal imaging and image processing of turbulent jet. Infrared, 33(5):42-47.

Gonzalez R C, Woods R E, Eddins S L. 2005. Digital Image Processing. Beijing: Publishing House of Electronics Industry.

Grinzato E, Marinetti S, Bison P G, et al. 2004. Comparison of ultrasonic velocity and IR thermography for the characterization of stones. Infrared Phys. Technol., 46:63-68.

He M C. 2011. Physical modeling of an underground roadway excavation in geologically 45° inclined rock using infrared thermography. Eng. Geo., 121(3-4):165-176.

He M C, Gong W L, Li D J, et al. 2009. Physical modeling of failure process of the excavation in horizontal strata based on IR thermography. International Journal of Rock Mechanics and Sciences, 19:0689-0698.

He M C, Gong W L, Zhai H M, et al. 2010a. Physical modeling of deep ground excavation in geologically horizontally strata based on infrared thermography. Tunnelling and Underground Space Technology, 25:366-376.

He M C, Jia X N, Gong W L, et al. 2010b. Technical note: Physical modeling of an underground roadway excavation vertically stratified rock using infrared thermography. International Journal of Rock Mechanics and Sciences, 47:1212-1221.

Jeng F C, Woods J W. 1991. Compound Gauss-Markov random fields for image estimation. IEEE Trans. Signal Process, 39:683-697.

Luong M P. 1995. Infrared thermographic scanning of fatigue in metals. Nuclear Eng. Design, 158:363-376.

Luong M P. 1990. Infrared thermovision of damage processes in concrete and rock. Eng. Fracture Mech. ,35:127-135.

Martin C D, Read R S, Martino J B. 1997. Observations of brittle failure around a circular test tunnel. International Journal of Rock Mechanics and Sciences, 34(7):1065-1073.

Maxwell S C, Young R P, Read R S. 1998. A micro-velocity tool to assess the excavation damaged zone. International Journal of Rock Mechanics and Sciences, 35(2):235-247.

Pastor M L, Balandraud X, Grédiac M, et al. 2008. Applying infrared thermography to study the heating of 2024-T3 aluminum specimens under fatigue loading. Infrared Phys. Technol, 51: 505-515.

Read R S. 2004. 20 years of excavation response studies at AECL's underground Research Laboratory. International Journal of Rock Mechanics and Sciences, 41(8):1251-1275.

Read R S, Chandler N A, Dzik E J. 1998. In situ strength criteria for tunnel design in highly-stressed rock masses. International Journal of Rock Mechanics and Sciences,35(3):261-278.

Sagong M, Bobet A. 2002. Coalescence of multiple flaws in a rock-model material in uniaxial compression. International Journal of Rock Mechanics and Sciences, 39:229-241.

Soille P. 2008. Morphological Image Analysis-Principles and Applications. 2nd edition. Beijing: Tsinghua University Press.

Spampinato L, Calvari S, Oppenheimer C, et al. 2011. Volcano surveillance using infrared cameras. Earth Sci. Reviews, 106:63-91.

Suganthi S S, Ramakrishnan S. 2014. Anisotropic diffusion filter based edge enhancement for segmentation of breast thermogram using level sets. Biomedical Signal Processing and Control,10:128-1236.

Tang X W, Ding H S, Yuan Y E, et al. 2008. Morphological measurement of localized temperature increase amplitudes in breast infrared thermograms and its clinical application. Biomedical Signal Process Control,3:312-318.

Tiago B B, Aura C, Rita C F L, et al. 2013. Breast thermography from an image processing viewpoint: A survey. Signal Process,93:2785-2803.

Wu L X, Liu X J, Wu Y H, et al. 2002. Changes in infrared radiation with rock deformation. International Journal of Rock Mechanics and Sciences, 39:825-831.

Wu L X, Wang J Z. 1998. Infrared radiation features of coal and rocks under loading. International Journal of Rock Mechanics and Sciences, 35(7):969-976.

Zeng M, Li J, Peng Z. 2006. The design of top-hat morphological filter and application to infrared target detection. Infrared Phys. Technol,48:67-76.

Appendix

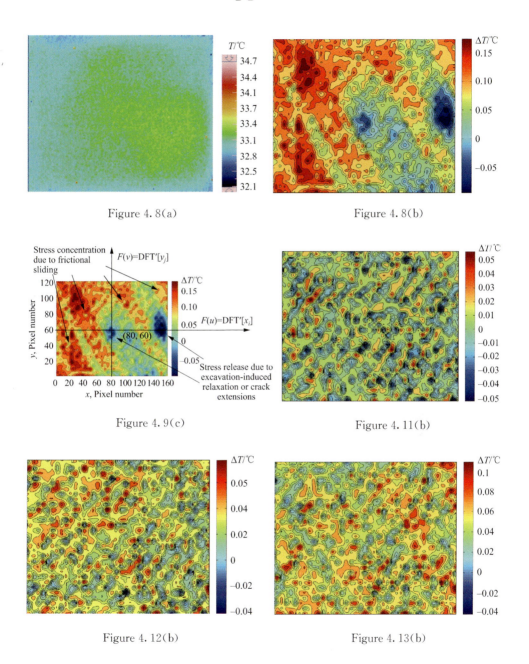

Figure 4.8(a)

Figure 4.8(b)

Figure 4.9(c)

Figure 4.11(b)

Figure 4.12(b)

Figure 4.13(b)

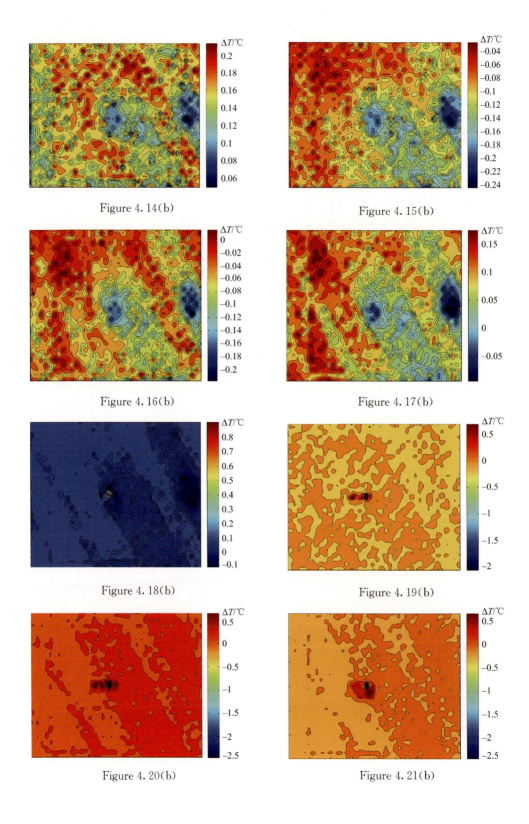

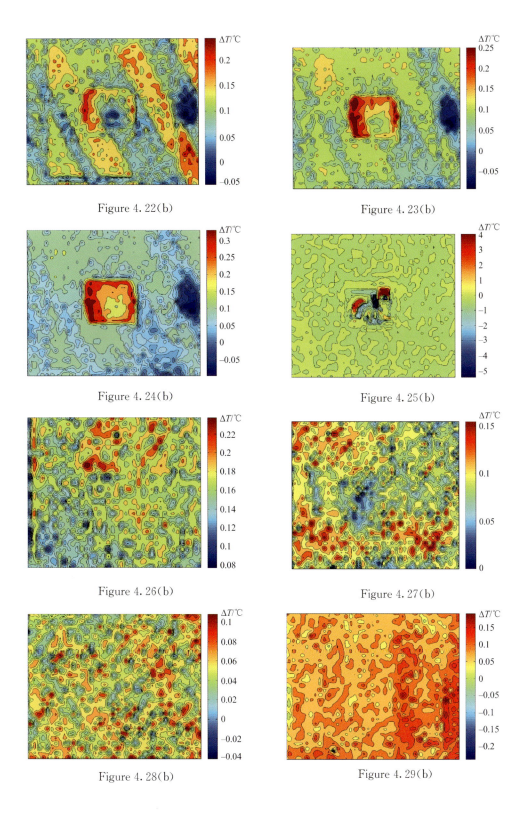

Figure 4.22(b)

Figure 4.23(b)

Figure 4.24(b)

Figure 4.25(b)

Figure 4.26(b)

Figure 4.27(b)

Figure 4.28(b)

Figure 4.29(b)

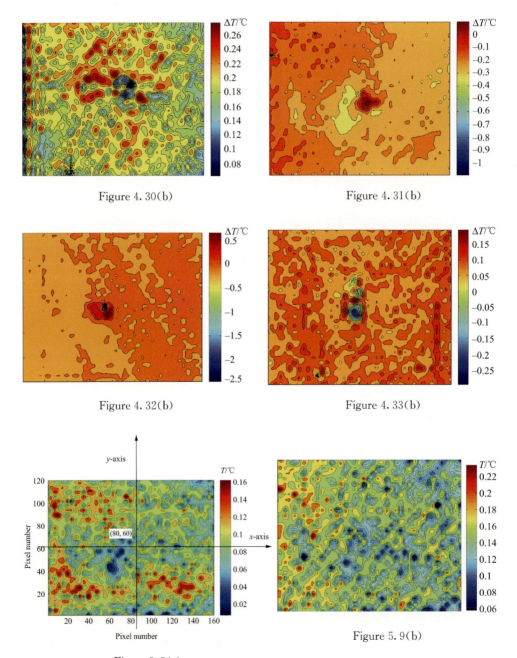

Figure 4.30(b)

Figure 4.31(b)

Figure 4.32(b)

Figure 4.33(b)

Figure 5.5(a)

Figure 5.9(b)

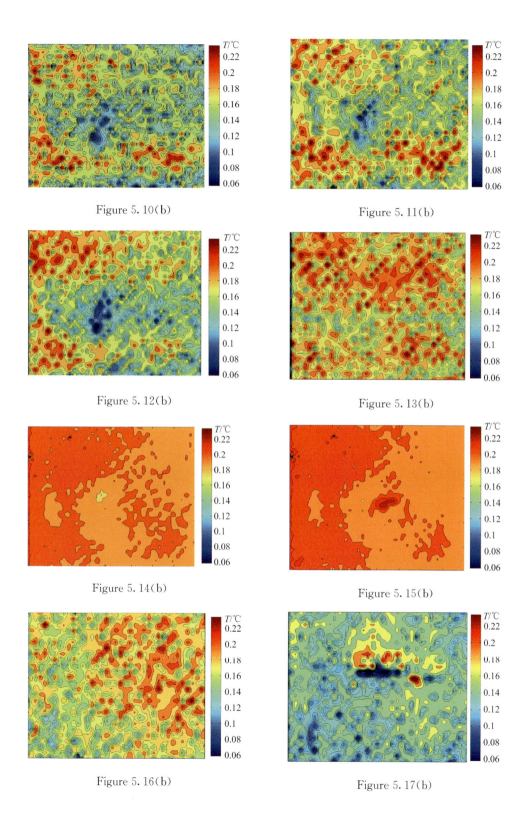

Figure 5.10(b)

Figure 5.11(b)

Figure 5.12(b)

Figure 5.13(b)

Figure 5.14(b)

Figure 5.15(b)

Figure 5.16(b)

Figure 5.17(b)

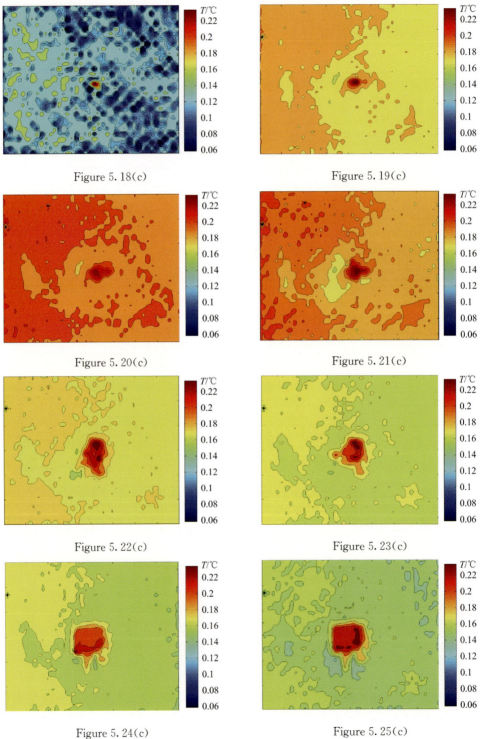

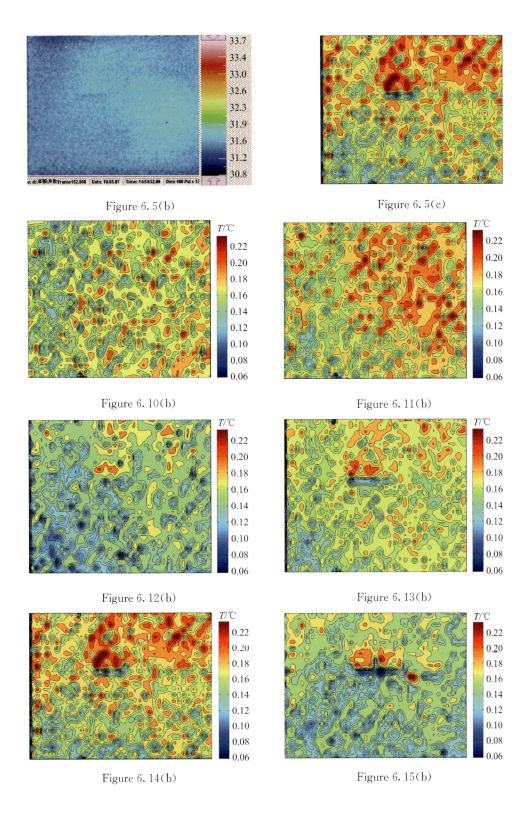

Figure 6.5(b)

Figure 6.5(c)

Figure 6.10(b)

Figure 6.11(b)

Figure 6.12(b)

Figure 6.13(b)

Figure 6.14(b)

Figure 6.15(b)

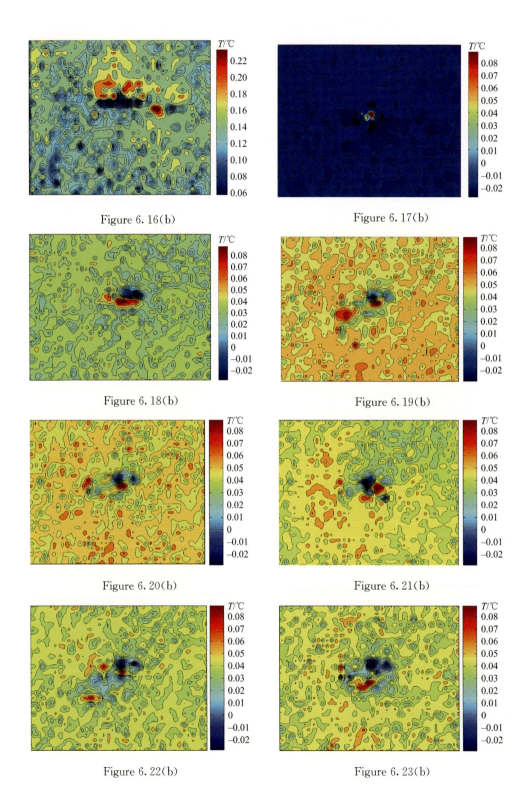

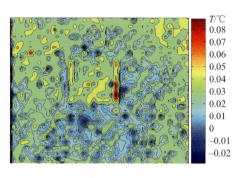

Figure 6.24(b)

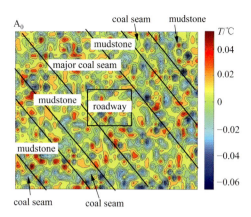

Figure 7.10

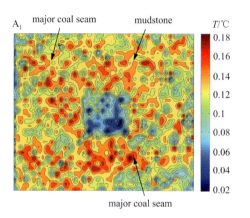

Figure 7.11

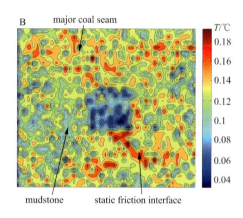

Figure 7.13

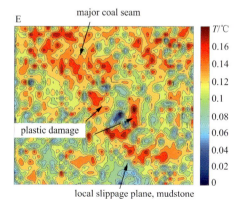

Figure 7.14

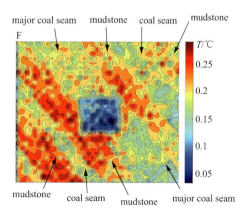

Figure 7.15

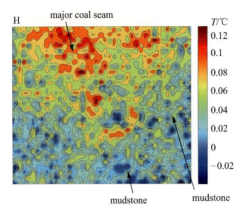

Figure 7.16

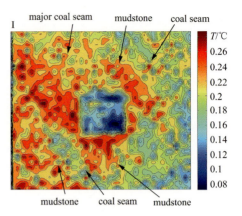

Figure 7.17

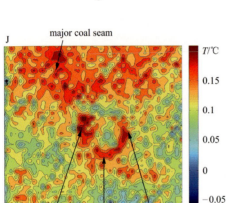

Figure 7.18

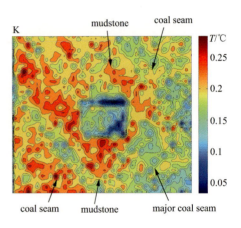

Figure 7.19

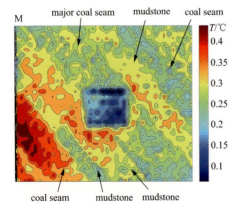

Figure 7.20

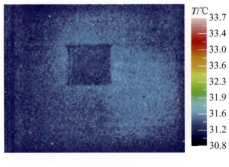

Figure 8.4(a)

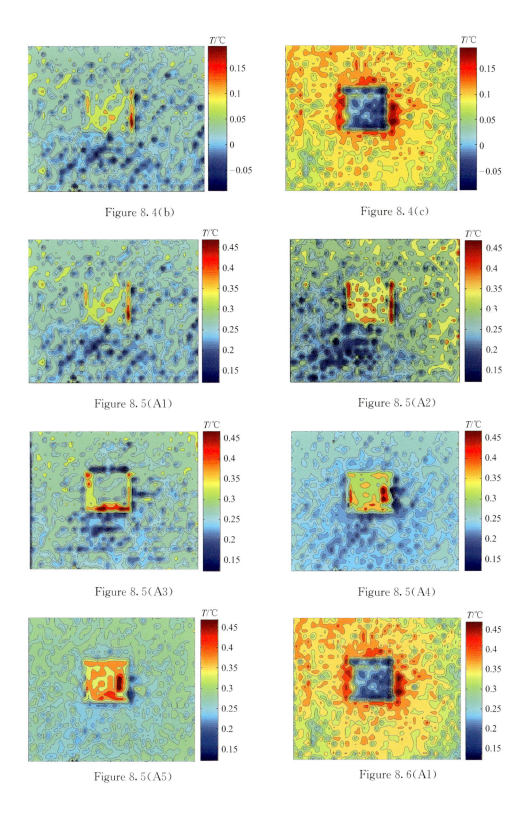

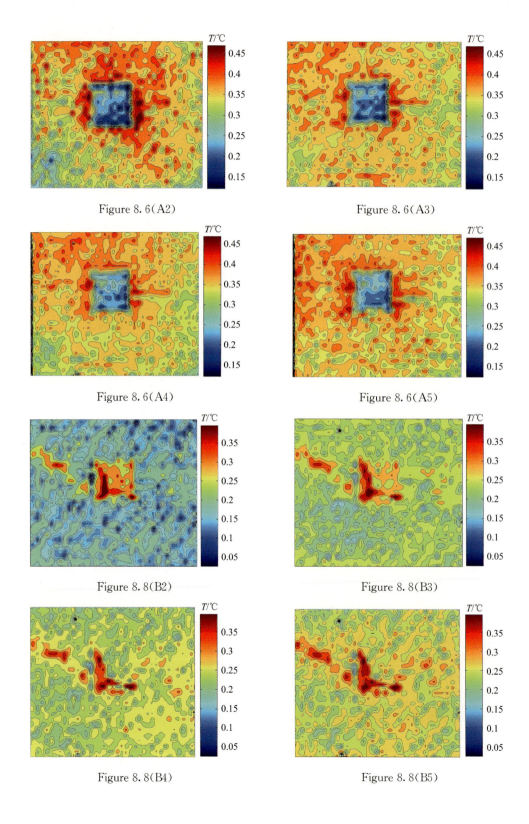

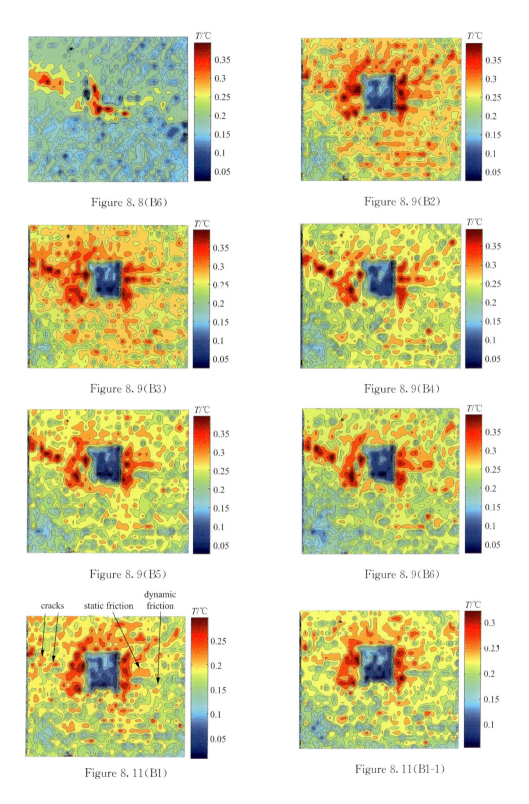

Figure 8.8(B6)

Figure 8.9(B2)

Figure 8.9(B3)

Figure 8.9(B4)

Figure 8.9(B5)

Figure 8.9(B6)

Figure 8.11(B1)

Figure 8.11(B1-1)

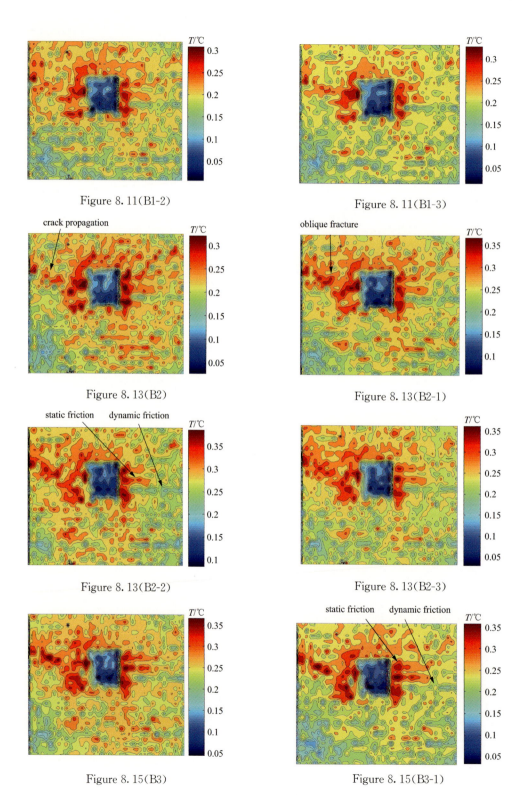

Figure 8.11(B1-2)

Figure 8.11(B1-3)

Figure 8.13(B2)

Figure 8.13(B2-1)

Figure 8.13(B2-2)

Figure 8.13(B2-3)

Figure 8.15(B3)

Figure 8.15(B3-1)

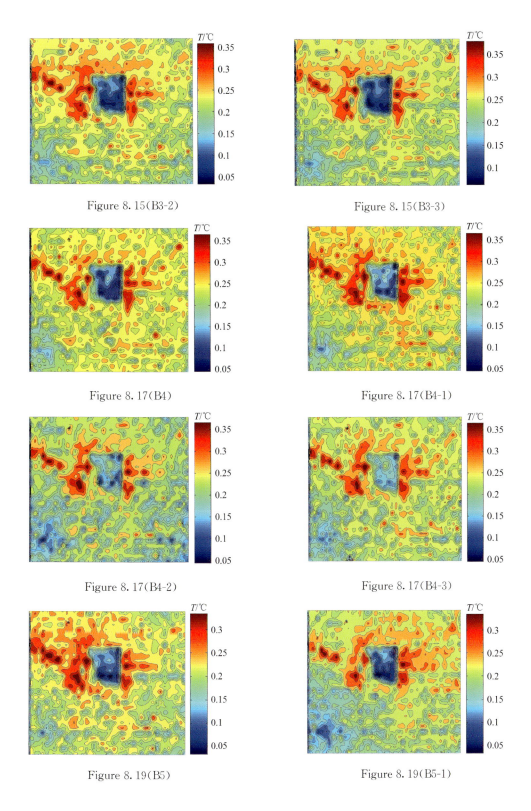

Figure 8.15(B3-2)　　　　　　　　　　Figure 8.15(B3-3)

Figure 8.17(B4)　　　　　　　　　　Figure 8.17(B4-1)

Figure 8.17(B4-2)　　　　　　　　　　Figure 8.17(B4-3)

Figure 8.19(B5)　　　　　　　　　　Figure 8.19(B5-1)

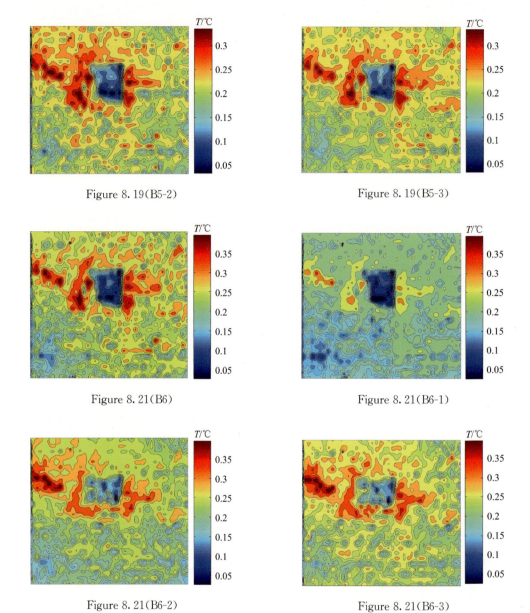

Figure 8.19(B5-2) Figure 8.19(B5-3)

Figure 8.21(B6) Figure 8.21(B6-1)

Figure 8.21(B6-2) Figure 8.21(B6-3)